Eye and Brain

Eye and Brain

The Psychology of Seeing

Fifth edition

by

RICHARD L. GREGORY

PRINCETON UNIVERSITY PRESS

PRINCETON, NEW JERSEY

Published by Princeton University Press,
41 William Street, Princeton, New Jersey 08540

This edition has been authorized by the Oxford University Press for
sale in the USA and Canada only and not for export therefrom

Library of Congress Cataloging-in-Publication Data
Gregory, R. L. (Richard Langton)
Eye and brain: the psychology of seeing / Richard L. Gregory—5th ed.
p. cm.—(Princeton Science Library)
Includes bibliographical references and index.
ISBN 0-691-04840-1
ISBN 0-691-04837-1 (pbk.)
1. Visual perception.   I. Title.   II. Series.
[DNLM: 1. Vision.   2. Visual Perception.   WW 103 G823e 1979]
BF241.G7   1990   89-72211
152.14—dc20   DNLM/DLC

Fourth edition, for the Princeton Science Library, 1990
Fifth edition 1997

Typeset by Footnote Graphics, Warminster, Wilts

Printed in Hong Kong

1 3 5 7 9 10 8 6 4 2

Contents

Pretext   vii

1 Visions of vision   1

2 Light   14

3 Eye   24

4 Brain   67

5 Seeing brightness   84

6 Seeing movement   98

7 Seeing colours   121

8 Learning how to see   136

9 Realities of art   170

10 Illusions   194

11 Speculations   244

Bibliography and notes   256

Index   269

Pretext

This greatly rewritten edition of my first book started from the happy experience of giving lectures and running practical classes (very important!) at the University of Cambridge, England. It was a privilege to share the excitement of trying to explain what we see and how we see, including the strange phenomena of illusions, with generations of students. Many are now friends and colleagues, continuing to be fascinated by the gradually revealed wonders of eye and brain.

This book was not written for examinations or for any formal teaching course, though it has by now been so accepted, especially for psychology, physiology, the visual arts (including architecture), physics, and philosophy. This should not be surprising for perception is the basis of all our experience and understanding, in science and art and everyday life. The study of perception is indeed central to what used to be called by the splendid name—and still is in Scotland—'Experimental philosophy'. For questions and speculation enliven all interesting science.

This book is an *introduction* to the psychology of vision. It is written to be read easily, and to be enjoyed; but this does not mean that its subject is easy or that I have glossed over difficult issues. To take almost the first page: one has to think quite hard just why babies do not have to learn to see things the right way up, though the images in their eyes are upside down. (And why do we see ourselves right–left reversed in a mirror, yet not upside down?)

This is an introduction for solo take-off to reflect about how we see. No one else can altogether see or think for us. I can only hope that this book will be useful and entertaining.

The question '**How do we see?**' may be approached from many points of view. Having approached it in one way we may, through change of insight, come to see it very differently. A classical account is that perception is the passive pick-up of information from the world, the brain having rather little to do. *Eye and brain* takes the very different view, that the brain (or mind) is highly active—constructing perceptions from hardly adequate information from the senses. On this view, illusions of many kinds take on remarkable significance, as phenomena well worth studying and trying to understand. Illusions have generally been written off as annoying and sometimes dangerous, but essentially trivial. A main theme of this book is to explain such phenomena of vision—which bridge art and science—as a way of discovering quite a lot about how perception works. It turns out that there are several very different kinds of illusions. Some are due to upsets of the physiology of the nervous system; others, very differently, are like incorrect hypotheses in science—due to inappropriate assumptions, or misplaced knowledge. The first kind of illusions may be compared with computer hardware errors; the second kind with bugs of software—though it does not follow that the brain is just like a digital computer.

How similar perception by machines is, or ever will be, to our own, is a topic of ever-growing interest as this new technology advances. In it, we see modern technology linked to ancient questions of philosophy—perhaps to find solutions even to why we are conscious.

Eye and brain first appeared in 1966, as the first volume of an imaginative series, World University Library, conceived by the distinguished London publisher Lord Weidenfeld. Each volume was lavishly illustrated, and translated into a dozen languages. This book owes a great deal to the original artists, Audrey Besterman and Mary Waldron, who drew so intelligently from my back-of-an-envelope doodles. Almost all of these first pictures are retained in this present much enlarged edition, together with others from *The intelligent eye* (1970), which is now out of print, and a large number of new pictures.

The second, third, and fourth editions (1972, 1977, 1990) of this book added new discoveries and ideas, though its structure remained essentially unchanged. The last 20 years have seen rapid growth of research. The brain sciences, including many kinds of studies of perception, have become a major international scientific endeavour, which has captured some of the greatest scientists from other fields— notably Francis Crick, who with James Watson and colleagues transformed how we think of life itself, through the discovery of the structure and significance of the DNA molecule.

This new edition is rewritten and greatly expanded. I hope it remains readable. It is, indeed, a considerable worry to tamper with a book that has been unusually successful over 30 years. It is a curious, and enjoyable experience, to criticize one's much earlier self and try to make use of the added experience and new ideas of one's later life.

Phenomena of illusion continue to be a major theme. Here is a new attempt to make sense of them, with a suggested classification. For any science, classification is extremely important. It seems high time to classify visual phenomena, by appearances and according to theoretical understanding, and this should be stimulated by seeing connections and differences more clearly. This is very much in the same spirit as Mendeleev's periodic table of the elements.

There is now, in this book, a lengthy discussion on the new work on seeing what babies see, and what and how they learn. This concludes with a short section on forgetting how to see: visual agnosia.

Studies of how motion is seen have always been important, but have now greatly advanced with the introduction of computer graphics. Unfortunately these phenomena cannot be demonstrated in a book. I would like to have included more on these experiments, particularly of Stuart Anstis and V. S. Ramachandran at UCSD in California; but perhaps these will have to wait for a new medium. We have, however, added 3-D red–green stereo, so that previously hidden phenomena may now be seen.

Attempts to give vision to machines remain of great interest, with a recent change of emphasis from digital computers to analogue processors, especially interactive neural nets. This is still in a state of flux, and it remains unclear just how far electronics can encapsulate visual brain function. As this is inherently technical, and has not yet yielded quite the dramatic results that were hoped for some years ago, regretfully I have not given it much space here.

Over the last few years there has been a remarkable burst of interest on consciousness, especially how sensations (*qualia*) may be explained. It remains unclear how or why they are caused, or what, if anything, they do. Although none of us know the answers, a few hints are suggested at the end of the book. The study of perception has been ongoing for at least 2000 years, and has accelerated from the work of Helmholtz in the last century, to the use of entirely new techniques, including imaging functions of the brain, with PET and NMR scanning. These promise new understanding of how physiological processes give cognition—understanding and perception. We may hope that later editions of this book will be able to report more fully on these extraordinarily exciting developments, just appearing over the horizon of what we can see.

Plates

In the plate section at the back of this book there are red–green stereo pictures designed so that they appear three-dimensional when viewed through red–green spectacles (with the red filter on the left). This gives an interesting new perspective on some classic illusions.

To order red–green spectacles, please contact David Burder, 3-D Images Ltd, 31 The Chine, Grange Park, London N21 2EA.

1

Visions of vision

The eye is a simple optical instrument. With internal images projected from objects in the outside world, it is Plato's cave with a lens. The brain is the engine of understanding. There is nothing closer to our intimate experiences, yet the brain is less understood and more mysterious than a distant star.

We have only to open our eyes, and spread before us lies a banquet of colours and shapes, shadows, and textures: a pageant of rewarding and threatening objects, miraculously captured by sight. All this, from two tiny distorted upside-down patterns of light in the eyes. Seeing is so familiar, apparently so easy, it takes a leap of imagination to appreciate that the eyes set extremely difficult problems for the brain to solve for seeing to be possible. How does it work? How are ghostly images transformed into appearance of solid objects, lying in an outer world of space and time?

From the beginnings of recorded questioning there have been several approaches to how we see. These are very different from current views. An essential problem was how distant objects reached eye and brain, while remaining out there in space. Two and half millennia ago, Greek philosophers thought that light shoots out of the eyes, to touch objects as probing fingers. A different notion at that time was that objects have expanding 'shells' like ripples from a stone dropped on a pool, but maintaining the object's shape to great distances. Called 'sense data' until quite recently by philosophers, they were supposed to be intermediaries—neither matter nor mind—between objects and perceptions. Both of these ideas were serious candidates before it was realized that in the eyes there are images of light, optically projected from the outside world onto the screens of the retinas. Optical images were unknown before the tenth century, and not until the start of the seventeenth were images discovered in eyes. At last it became clear that light does not enter or leave the brain, locked privily in its box of bone. All the brain receives are minute electrochemical pulses of various frequencies, as signals from the senses. The signals must be read by rules and knowledge to make sense. Yet what we *see*, and what we *know*, or believe, can be very different. As science advances, differences between perceived appearances and accepted realities become ever greater.

This is far beyond the common account that the eye is a camera; yet this is essentially true, though far from the whole story. It is the uncamera-like features of eyes and brains that most interest us here.

What is striking is the huge amount of brain contributing to vision, giving immense added value to the images of the eyes. Where does this extra richness for vision come from? By some authorities it is simply denied—they see perception as passive acceptance of what is out there, as a window facing the world. But this does not begin to explain how we see objects from the sketchy images of the eyes, even from sparse lines and crude dots of seemingly inadequate pictures. In ideal conditions, object perception is far richer than any possible images in the eyes. The added value must come from dynamic brain processes, employing knowledge stored from the past, to see the present and predict the immediate future. Prediction has immense survival value. It not only makes fast games possible in spite of the physiological signal delays from eye to brain, and brain to hand. Anticipating dangers and potential rewards is essential for survival— made possible by buying time from seeing objects distant in space.

This introduces a particular kind of way of thinking about perception. It is essentially the view of the nineteenth century German polymath—physiologist, physicist, psychologist—Hermann von Helmholtz (1821–1894) who described perceptions as 'unconscious inferences' from sensory data to what might be out there. This is the 'school of thought' accepted here, but there are others. Psychology is unusual among the sciences, in having doubts of its most basic assumptions, with very different alternatives held by different authorities. 'Active' and 'passive' accounts are extremely different ways of describing and explaining phenomena of vision. Few other sciences have such divisions in their basic ideas. There were equally dividing paradigms (as the American philosopher of science Thomas Kuhn calls them) at the time of Darwin, over whether there was evolution of species by natural selection or special creation of each species, but this is now resolved, by almost universal acceptance of evolution. However, although active, essentially Helmholtzian, accounts of perception are now dominant, this was not so a few years ago, and they are not universally held today.

Paradigms of perception

It might be useful to outline some recent kinds of explanations:

Behaviourism was founded by John Broadus Watson (1878–1958) with his manifesto of 1913: 'Psychology as the behaviourist views it'. This set out to deny consciousness, at least as a ploy to make psychology scientifically respectable. Behaviourism was extremely influential in America until the 1980s, especially with the experiments and ideas of B. F. Skinner. It is based on the earlier work of the Russian physiologist Ivan Petrovich Pavlov (1849–1936) with his experiments on conditioned (or 'conditional') reflexes. Pavlov showed that, starting with an innate (inherited) reflex, such as salivating to the sight or smell of meat, dogs would come to salivate to any stimulus (such as a bell) presented at the same time or just before the food. It proved possible to build up chains of conditioned reflexes. For the behaviourists, it seemed that chains of conditioning would explain all learned behaviour, even language.

They listed innate reflexes observed in babies, and measured strengths of drives for rewards. So they developed a scientifically respectable-looking 'atomism' for describing complex behaviour from simple components.

Problem solving, at least for animals such as cats, was supposed to be by trial-and-error—without insight into the nature of a problem. Perception and behaviour were supposed to be controlled quite directly by stimuli, with modifications from internal states of drives such as hunger, so that with sufficient knowledge psychology should become a perfectly predictive science. This has not worked out. Watson's denial of consciousness makes psychology even more like physics; but for most of us today, it **threw the baby out with the bath water**.

Gestalt psychology was a very different rival school. Founded by a group of German scientists in the 1920s, the emphasis was on dynamics and 'holism'. Many of the Gestalt psychologists (entirely different from recent Gestalt therapy) fled from Nazi Germany under Hitler, to settle in America where they had a major influence.

A 'Gestalt' was a grouping of elements such that the whole is greater than the sum of its parts. Analysis into perceptual components was not supposed to be possible. An important concept was 'pregnance', roughly, 'pregnant with meaning'. Problem solving was supposed to be by 'insight'. A famous example is Wolfgang Kohler's chimpanzee Sultan; presented with a banana out of reach and a number of short sticks, he was described as looking at the sticks for several minutes, then suddenly joining two together to pull down the banana. Throughout there is an emphasis on sudden solutions, with sudden insights.

The Gestalt psychologists described visual perceptions as more than the sum of stimuli, organized according to various laws. These were mainly derived from subjective reports of how arrangements of dots are seen as patterns: which dots 'belong' together, form lines and so on, or are separate. This may seem vague, hardly 'scientific'; but the Gestalt laws of organization have turned out to be important for perception of sight and sound. They have been taken up by the artificial intelligence (AI) community, especially for programming computers to recognize patterns and objects. The laws include:

(1) closure—tendency for a roughly circular patterns of dots to be seen as 'belonging' to and forming an object;

(2) common fate—parts moving together, as leaves of a tree, seen as an object;

(3) contiguity of close-together features; and a preference for smooth curves.

The laws of organization were supposed to be inherited, but as they correspond to common features of almost all objects, learning could be involved, to give us all much the same visual organizations. Gestalt notions of brain physiology (electrical fields and so on) have been abandoned.

Cognitive psychology, in its various forms, denies that perception and behaviour are controlled by stimuli, emphasizing the importance of general background knowledge and more-or-less logical thought processes. How far these apply to perception is controversial. Generally, visual perception has been thought of as quite separate from cognitive problem solving, but this can be questioned. However, Hermann von Helmholtz did think of visual perceptions as unconscious inferences, and so related perception to thinking.

The Cambridge psychologist Kenneth Craik (1914–45) put forward the notion that the brain works with physiologically existing functional 'internal models' of perceived and imagined objects and situations. This is now generally modified to a more symbolic account; but the notion of *representing* by the brain is accepted as central to cognitive approaches.

The intelligent eye

This philosophy, or paradigm, is largely derived from Helmholtz. It is, that visual and other perception is intelligent decision-taking, from limited sensory evidence. The essential point is that sensory signals are not adequate for direct or certain perceptions; so intelligent guessing is needed for seeing objects. The view taken here is that perceptions are predictive, never entirely certain, *hypotheses* of what may be out there.

It was, perhaps, the active intelligence of perception that was the evolutionary start of conceptual problem-solving intelligence. When, a generation before Freud, Helmholtz called perceptions unconscious inferences he was much criticized—for how could blame or praise be applied to unconscious perceptions, and actions? We are still puzzled by these issues.

There are many traps along the way of exploring Eye and Brain. It is important to avoid the temptation of thinking that eyes produce pictures in the brain which are perceptions of objects. The pictures-in-the-brain notion suggests an internal eye to see them. But this would need a further eye to see *its* picture—another picture, another eye—and so on forever, without getting anywhere. Early this century, the Gestalt psychologists held that perceptions were pictures inside the brain: supposed electrical brain fields copying forms of objects. So a circular object would produce a circular brain field. Presumably a house would have a house-shaped electrical brain-picture, though this is far less plausible. A green object having a green brain-trace is ridiculous. This notion, known as isomorphism, led to supposing that properties of brain fields produce visual distortions (like bubbles tending to be spherical), visual phenomena being explained by their supposed mechanical or electrical properties. There is no evidence for isomorphic brain traces.

We now think of the brain as *representing*, rather as the symbols of language represent characteristics of things, although the shapes and sounds of language are quite different from whatever is being represented. Language requires *rules* of grammar (syntax), and *meanings* of symbols (semantics). Both seem necessary for processes of vision; though its syntax and semantics are implicit, to be discovered by experiment.

Some puzzles of vision disappear with a little thought. It is no special problem that the eyes' images are upside down and optically right–left reversed—for they are not seen, as pictures, by an inner eye.

As the image is not an object of perception, it does not matter that it is inverted. The brain's task is not to see retinal images, but to relate signals from the eyes to objects of the external world, as essentially known by touch. Exploratory touch is very important for vision. It matters that touch–vision relations remain unchanged. When changed experimentally (with optically reversing prisms, or lenses or mirrors) then a problem is set up, and special learning is required. No special learning is needed for a baby to see the world the right way up.

The Gestalt psychologists made some excellent suggestions, realizing that the visual system has to solve some very difficult problems, which arise right at the start. How does the mosaic of retinal stimulation give perception of individual objects? (This also applies to hearing, especially of speech. We can only distinguish separate words in familiar languages). The visual separations of objects are not given simply by borders of light on the retinas. Separations into objects is given by various rules, and by knowledge. Sharp borders are rather rare, except for line drawings, which are not typical and present their own problems.

The tendency to group elements into wholes was investigated by the Gestalt psychologists, with patterns of dots. Their experiments suggested various rules of organization for object creation. This was a much more useful idea than isomophism, and it has turned out to be important for programming visual computers, though this development is in its infancy. We can see something of dynamic grouping in an array of dots (see Figure 1.1). The dots are equally spaced, but there is a tendency to 'organize' them into columns and rows. We can see in ourselves active try-outs for organizing visual data into objects. With more complexity, groupings and re-groupings can become more dramatic (Figure 1.2).

If the brain were not continually trying out organizations of data, for searching for objects, such as faces, the cartoonist would have a hard time. In fact all he or she has to do is to present a few well-chosen lines and we see a face, complete with an expression. This essential process of vision can, however, go over the top to make us see faces in the fire, galleons in the clouds, or the Man in the Moon. Vision is certainly not infallible. This is largely because knowledge and assumptions add so much that vision is not directly related to the eyes' images or limited by them—so quite often it produces fictions. This can be useful, as images are inherently inadequate, but visual fictions, and other illusions, worry philosophers seeking certainty from sight.

How did such complex processes for representing things start?

1.1 These equally spaced dots are seen as continually changing patterns of rows and squares. We see something of the dynamic organizing power of the visual system.

What is their evolutionary benefit? For simple organisms, the eyes' signals do initiate behaviour quite directly. Thus tropisms toward or away from light may serve to find protection or food in typical conditions, without the creature being capable of making decisions between alternative courses of action. We might say that primitive organisms are almost entirely controlled—tyrannized—by tropisms and reflexes.

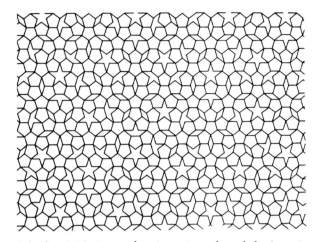

1.2 These circles form intricate ever-changing patterns through the dynamics of vision.

Many reflexes still protect us (such as blinking to a puff of air on the eye, or to a sudden loud sound) and reflexes are essential for the maintenance of body functions such as breathing and digestion. But gradually, through evolution, direct control from outside objects has been largely replaced by more and more indirect representations of objects and situations. This has the huge advantage that behaviour can be appropriate to properties of objects that are not and often cannot be signalled by the senses. Thus we pick up a glass to drink not simply from stimuli, but from knowledge of glasses, and what they may contain. By contrast, a frog surrounded by dead flies will starve to death, for though they are edible it does not see them, as they do not move.

These brain representations are far more than pictures. They include information of what various kinds of objects may do, or be used for. For behaviour to be appropriate in a wide variety of situations requires a great deal of knowledge of the world. Knowledge must be selected and accessed within a fraction of a second to be useful for perception, or the moment for action (or survival) will pass. So the intelligence of vision works much faster than other problem solving. This may be why perceptions are quite surprisingly separate from generally more abstract conceptions, and may disagree. Thus, one experiences an illusion, though one knows it is an illusion and even what causes it. Illusions tell us a great deal—sometimes, as I shall show, more than we would wish to know!

Here, I have introduced the kind of approach to vision which is developed in this book. This may be called an *indirect* and *active* account. Not all authorities will agree with it. The alternative—that perceptions are directly from the external world—was argued most strongly by the American psychologist James J. Gibson (1904–1979), at Cornell University. His experimental work, especially on moving and stationary gradients for depth perception (Figure 9.18), is justly celebrated. Gibson considered that seen depth and form are determined by patterns such as these, from what he calls the 'ambient optical array' of light. Gibson's theory is a kind of realism, in which perceptions are supposed to be 'picked up' from the world, rather than created as representations. Realism has always been attractive to philosophers, as it promises reliable perception, and so unquestionable bases for empirical knowledge. Given retinal images, and the complexities of the physiology of vision, as well as the richness of illusion (see Chapter 10), it is hard to see how this can be literally true. Gibson's essentially *passive* account is very different from the notion in this book, that perceptions are constructed hypotheses.

Phenomena of illusion are played down as embarrassments by *direct* theorists such as Gibson, but are grist to the mill, and evidence for, knowledge-based processes of perception. But if past experience, assumptions, and active processing are important, there can hardly be raw data for vision. Rather, we might say that perceptual data are cooked, by processes we shall look at here in considerable detail.

Perceptions as hypotheses

The indirectness of vision and its complexity are evident in its physiology. As much as half the cortex of the brain is involved, including rich 'down-going' pathways. Its activity is evident from physiological recordings and from many visual phenomena. These include ambiguities of various kinds, and our ability to see objects in a few sketchy lines of a picture.

Cartoons (such as Figure 1.3) bring out the importance and power of knowledge and assumptions for seeing more than meets the eye.

There are many well known ambiguous figures, which show that the same pattern of stimulation at the eyes can give rise to very different perceptions. There are three kinds (Figure 1.4): those which alternate as *objects* or *space* between objects (figure–ground); those which spontaneously switch in *depth*; and those which change from one *object* into a different object or kind of object.

We shall have a lot to say about visual ambiguities as these phenomena are intriguing and useful for teasing out kinds of processes of

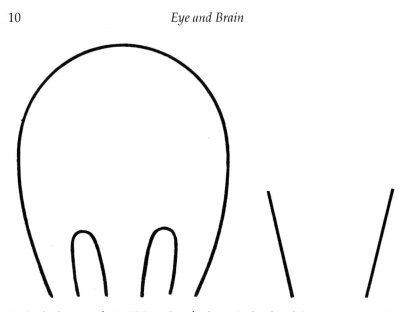

1.3 A joke figure—what is it? Just a bunch of meaningless lines? As soon as you see it as objects, it will suddenly appear almost solid, as objects not mere patterns. (The answer is given at the end of this chapter.)

vision. They are important in art, and they can present hazards in real-life situations such as driving and flying. The essential problem for the brain to solve is that any given retinal image could be produced by an infinity of sizes and shapes and distances of objects, yet normally we see just one stable object. So the usual lack of phenomenal ambiguity is even more remarkable than the brain's occasional failure to make up its mind.

A central notion here is that perceptions are *hypotheses*. This is suggested by the fact that retinal images are open to an infinity of interpretations, and from the observed phenomena of ambiguity. The notion is that perceptions are like the predictive hypotheses of science. Hypotheses of perception and of science are risky, as they are predictive and they go beyond sensed evidence to hidden properties and to the future. For perception, as for science, both kinds of prediction are vitally important because the eye's images are almost useless for behaviour until they are read in terms of significant properties of objects, and because survival depends on behaviour being appropriate to the immediate future, with no delay, although eye and brain take time to respond to the present. We behave to the present by anticipation of what is likely to happen, rather than from immediate stimuli.

Seeing a table, what the eye actually receives is a grainy pattern

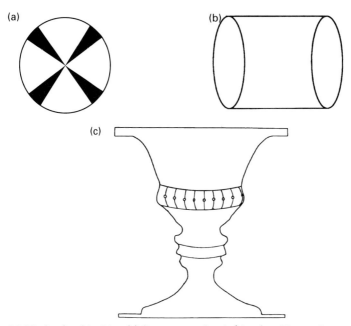

1.4 Kinds of ambiguities: (a) figure–ground switching (or object and space swapping), (b) depth flipping, (c) object changing.

which is read as wood—though it might be a plastic imitation, or perhaps a picture. Once the *wood* hypothesis is selected, behaviour is set up appropriately. People who live with plastic tables pretending to be wood do terrible damage to mahogany! Science and perception work by knowledge and rules, and by analogies. Figure 1.5 gives an archaeological analogy. Some of the holes in the ground might be ancient post holes; others might be rabbit holes, to be ignored. One group of archaeologists accepted close-together large holes as evidence for a grand entrance. These were altogether rejected by the other archaeologists. One group constructed a large rectangular hut: the other, a small rectangular hut and a circular building. 'Bottom-up' rules—holes being close together and forming straight lines or smooth curves, and 'top-down' knowledge or assumptions of what kinds of building were likely—affected the 'perceptions'. Both could have been wrong.

The visual brain has the same kind of problem for accepting or rejecting evidence from patterns of photons in the eyes. Seeing objects involves general rules, and knowledge of objects from previous experience, derived largely from active hands-on exploration.

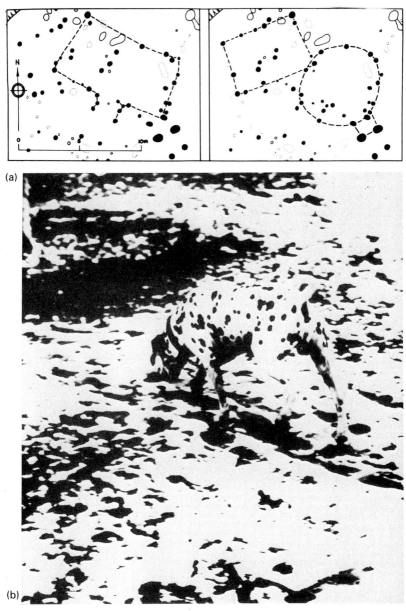

(a)

(b)

1.5 Which hypothesis is right? (a) Two groups of archaeologists postulated very different ancient buildings from the same holes in the ground, by using general rules for accepting or rejecting holes as genuine, and knowledge of likely kinds of buildings. Perhaps neither solution is correct. (b) A Dalmatian dog. One has to guess much of the outlines from the Dalmatian dog hypothesis.

A central theme of this book is that perceptions are much like hypotheses of science. Hypotheses of science do not only have ambiguities; they can also have or produce, distortions, paradoxes, or fictions. It is interesting that all of these can appear as phenomena of perception. For under various conditions which can be set up and investigated, as well as in normal life, we can see distortions, paradoxes and fictions. With ambiguities, these turn out to be key phenomena for investigating and understanding perception.

The big difference between hypotheses of science and perceptual hypotheses is that only perceptions have consciousness. We will hazard a guess as to why this might be at the end of the book.

What is so special about human vision? We learn a lot from observing other animals; but only humans can draw or paint representations, and only humans have a structured language. It turns out that both pictures and language depend on imaginative use of ambiguities.

Our early human ancestors were able to represent and see mammoths and bison in sketchy lines and blodges on cave walls. It was presentations of alternative realities, and playing with fantasies through ambiguities, that released mankind from the tyranny of reflexes and tropisms of our distant ancestors. So humans took off from nature, into evocative art and questioning science, making civilization possible.

Why we are so biologically special, with our huge brain and knowing eyes, is beyond the range of this book and the ability of its author to answer.

Key to Figure 1.3

Washer woman with pail

2

Light

To see we need light. This may seem too obvious to mention but it has not always been recognized. Greek philosophers, including Plato and Euclid, thought of vision as due not to light entering but rather to particles shooting out of the eyes, as fingers touching surrounding objects. It is hard to imagine now why they did not try to settle the matter with a few simple experiments. Although to philosophers the problem of how we see has always been a favourite topic of speculation, only in the last hundred and fifty years or so has it been subjected to systematic experiments. This is odd, for all observations in science depend upon the human senses, especially sight.

How are images produced? The simplest way an image can be formed is by a pinhole. Figure 2.1 shows how this comes about. A ray from a point of the object reaches a point on the screen by the straight line passing through the pinhole, so an upside-down and right–left reversed picture of the object is formed. The pinhole image will be dim, for the hole must be small for the image to be sharp. (Though if too small it will be blurred, because the wave structure of the light is upset.)

A lens may be thought of as a pair of curved prisms (Figure 2.2). It directs a lot of light from each point of the object to a corresponding point on the screen, thus giving a bright image. Lenses only work well when suitable and adjusted correctly.

What is light?

In the last three hundred years there have been two rival accounts of the nature of light. Sir Isaac Newton (1642–1727) thought of light as a shower of particles. The Dutch physicist Christiaan Huygens (1629–95; Figure 2.3) proposed that it is pulses travelling through an all-pervading medium—the *aether*—which he thought of as small elastic balls in

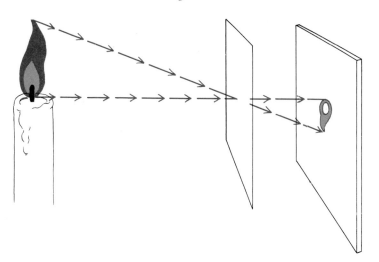

2.1 Forming an image with a pinhole. A ray from a given region of the source reaches only a single region of the screen by passing through the hole. Thus an inverted and sideways reversed image is built up from rays passing through the hole. The image is free of distortion, but it is dim and not very sharp. A very small hole gives a still dimmer image, with increased blurring through diffraction effects due to the wave nature of light.

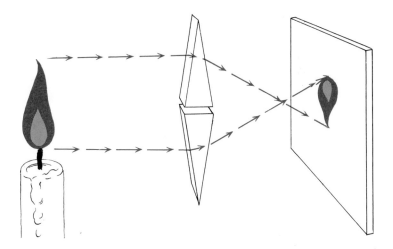

2.2 A focusing lens can be thought of as a pair of converging prisms. The image is far brighter than from a pinhole, but it is generally distorted in some degree, and the range of sharp focus is limited. (This figure should not be taken too literally—image-forming lenses have curved surfaces.)

2.3 Christiaan Huygens (1629–95) by an unknown artist. As well as being a highly creative scientist, Huygens had an unusually active life, as secretary to the Prince of Orange. He invented the pendulum clock (suggested by Galileo on his death bed) and showed how to make a strictly isochronous (equal time for any amplitude) pendulum. He correctly described Saturn's ring, and discovered (at the same time as Christopher Wren) the laws of elastic bodies. He argued that light travels as waves through an all-pervading *aether* of tightly packed balls.

contact with each other. Any disturbance, he suggested, would be carried in all directions through the packed spheres as a wave, and this wave is light.

The controversy over the nature of light is one of the most interesting in the history of science. A crucial question in the early stages of the discussion was whether light travels at a finite speed or whether it arrives instantaneously. This was answered in an entirely unexpected way by a Danish astronomer, Olaus Roemer (1644–1710). He was engaged in recording eclipses of the four bright satellites orbiting Jupiter, and found that the times he observed were not regular, but depended on the distance of Jupiter from the earth. He came to the conclusion (in 1675) that this was due to the changing time light took to reach him from Jupiter's satellites—the time increasing when the distance increased—because of the finite speed of light. The distance of Jupiter varies by about 300 000 000 km (twice the distance of the Sun from Earth), and the greatest time difference he observed was 16 minutes 36 seconds earlier or later than the calculated time of the eclipses of the satellites. From his somewhat faulty estimate of the distance of the Sun, he calculated the speed of light at 309 000 km per second. With our modern knowledge of the diameter of the Earth's orbit we correct this to about 300 000 km per second, or 3×10^{10} cm/s. The speed of light has since been measured very accurately over short distances on earth and is now regarded as a primary constant of the universe.

Because of the finite speed of light and, more important for terrestrial objects, the considerable delay while nervous messages reach the brain, we always sense the past. Our perception of the Sun is eight minutes late, and all we know of the furthest object visible to the unaided eye (the Andromeda nebula) is from before humans appeared on earth. For nearby objects, there is the neural delay of several hundredths of a second, which is significant for fast action.

The value of 3×10^{10} cm/s for the speed of light strictly holds only for a perfect vacuum. When light travels through glass, or water, or any other transparent substance, it is slowed down to a velocity which depends on the refractive index (roughly the density) of the medium through which it passes. This slowing down of light is extremely important. It is this that causes prisms to bend light and lenses to form images. The principle of refraction (the bending of light when passing through changes of refractive index) was first understood in 1621 by a Dutch mathematician at Leyden, Willebrod von Roijen Snell (1591–1626). Snell died at thirty-five, leaving his results unpublished. The French philosopher–mathematician René Descartes (1596–1650) published the law of refraction, or sine law, 11 years later, (Figure 2.4).

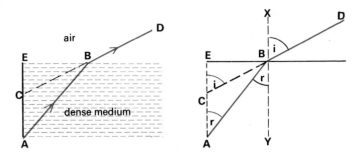

2.4 The sine law of refraction. When light passes from a medium A into a medium B, the sine of the angle of incidence bears to the sine of the angle of refraction a constant ratio.

We can see what happens with a simple diagram: if AB is a ray passing from a dense medium (glass) into a vacuum (or air) the ray will emerge into the air at some angle i along BD. The law states that $\sin i = u / \sin r$. The constant u is the refractive index of the glass, or other refracting substance.

Newton (Figure 2.5) thought of his corpuscles of light as being attracted to the surface of the denser medium, while Huygens thought that the bending was due to the light travelling more slowly. It was many years before the French physicist Jean Bernard Léon Foucault (1819–68) showed by direct measurement that light does indeed travel more slowly in a denser medium. It seemed for a time that Newton's corpuscle theory of light was entirely wrong—that light is purely a series of waves radiating through the *aether*—but at the beginning of the twentieth century it was shown that the wave account is not adequate for all the phenomena of light. It now seems that light is both particles and waves.

When light is bent by a prism, each wavelength (or frequency) is deviated through a slightly different angle, so the beam comes out as a fan of light, with the spectral colours spread out. Newton discovered that white light is a compound of the spectral colours by splitting a beam of sunlight into a spectrum in this way (Figure 2.6); then found that he could recombine the colours back into white light, by passing the spectrum through a second similar prism held the other way up. Newton named seven spectral colours: red, orange, yellow, green, blue, indigo, violet. One does not really see indigo as a separate colour, and orange is a bit doubtful; but Newton liked seven, as a magical number, and he thought of the spectrum in terms of the notes of a musical stave, so he added the names orange and indigo to make up seven.

We know now, though Newton did not, that each spectral colour or hue is light of a different frequency. We also know that all electro-

D. ISAACVS NEWTON EQVES
REG. SOCIETATIS PRÆSES AN.º 1703.

2.5 Sir Isaac Newton (1642–1727), by Charles Jervas. On the whole Newton held that light consists of particles, but he anticipated many of the difficulties which have been faced by the modern theory that light has dual properties of particles and waves. Newton devised the first experiments to show that white light is a mixture of the spectral colours, which paved the way to an understanding of colour vision. He realized that light itself is not coloured—but elicits colours—created as we now know by specialized cells of the brain.

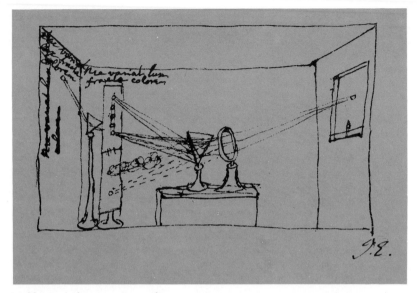

2.6 Newton's drawing of one of his experiments on colour. He first split light into a spectrum (with the large prism), then allowed light of a single colour to pass through a hole in a screen to a second prism. This did not produce more colours. He also found that a second prism placed in the spectrum would recombine the colours into white. Thus white light (sunlight) is made up of all the colours the spectrum.

magnetic radiation is essentially the same; the physical differences between radio waves, infrared, visible light, ultraviolet, X-rays, and so on are only their frequencies. A very narrow band of this huge range of frequencies, less than an octave in width, stimulates the eye to give vision. Figure 2.7 shows how this narrow window of the total spectrum fits into the physical picture. Looked at in this way we are almost blind!

From the speed of light (λ) it is a simple matter to calculate wavelength from frequency, or vice versa, as $F \lambda = 3 \times 10^8$ metres / second. In fact frequency is difficult to measure directly: it is easier to measure visible wavelengths, though this is not so for the much lower-frequency radio waves. The wavelength of light is measured by splitting it up with a grating of finely ruled lines, which also produces the colours of the spectrum. (This can be seen by holding a compact disc at an oblique angle to a source of light, when the reflection will be brilliantly coloured.) From the spacing of the lines of an optical grating, and the angle reflecting light of a given colour, wavelength may be determined very accurately. Blue light has a wavelength of about 4×10^{-7} m. Red light is about 7×10^{-7} m.

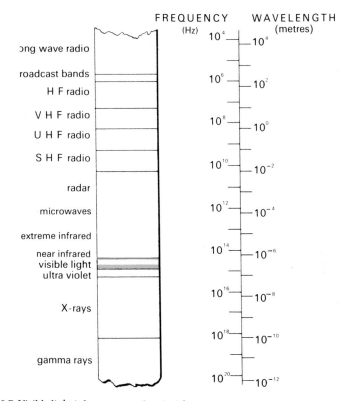

FREQUENCY
(Hz)

WAVELENGTH
(metres)

ɔng wave radio

roadcast bands

H F radio

V H F radio

U H F radio

S H F radio

radar

microwaves

extreme infrared

near infrared
visible light
ultra violet

X-rays

gamma rays

2.7 Visible light is but a narrow band of the total electromagnetic spectrum. The different properties of various frequencies are due to how electromagnetic radiations react with matter.

The range of wavelengths which is accepted for vision is important, for it sets the limit to the eyes' resolution, just as for optical instruments such as microscopes and cameras. Resolution increases with the frequency of the light and with the size (optical aperture) of an eye or instrument. Eyes are adapted to accept wavelengths of maximum energy of sunlight, without undue damage to the biological materials of which they are made. Compound eyes of insects work in the ultra-violet, no doubt because they are so small.

The present wave account of light is more complicated than ripples on a pond, for light is electromagnetic—meaning it is waves of both electrical and magnetic fields. These are at right angles to each other, and displaced by half a wave length, so that as one increases the other decreases (Figure 2.8).

Light is also thought of as consisting of particles of energy, called

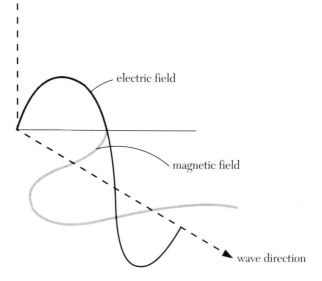

electric field

magnetic field

wave direction

2.8 Light is waves of electrical and magnetic fields, at right angles to each other, one increasing as the other decreases.

photons for visible light. *Quanta* is the general term used for all wavelengths or frequencies of radiation. These combine the characteristics of particles and of waves. Photons are sometimes thought of as packets of waves. The shorter the wavelength, the more waves in each photon. This is expressed by saying that the energy of a single photon (or quantum), is a function of frequency, such that $E = hv$, where E is the energy in ergs, h is a small constant (Planck's constant) and v is the frequency of the radiation.

We cannot, with the unaided eye, see individual quanta of light—known as 'photons'; but the receptors in the retina are so sensitive that they can be stimulated by a single photon, though something like five to eight are required for experience of a short dim flash. The individual receptors of the retina are as sensitive as it is possible for any light detector to be, since a quantum is the smallest amount of energy which can exist. It is rather sad that the transparent media of the eye do not quite match this development, as less than half the quanta reach the receptors, the rest being lost by absorption and scattering within the eye. In spite of this loss it would be possible, under ideal conditions, to see a single candle placed nearly twenty miles away.

The first experiment for measuring the number of quanta required for the eye to detect light was undertaken in 1942 by three physiologists, S. Hecht, S. Schlaer, and M. H. Pirenne. Realizing that the eye must be almost as sensitive as theoretically possible, they devised an ingenious experiment depending on probabilities, based on a function, the Poisson distribution. This gives the expected distribution of hits on a target. The idea is that at least part of the moment-to-moment variation in the effective sensitivity of the eye is not due to anything in the eye, or the nervous system, but to the moment-to-moment variation of energy of a weak light source when there are only a few quanta arriving. Imagine a desultory rain of bullets: they will not arrive at a constant rate, but will fluctuate. Similarly there is fluctuation in the number of light quanta arriving at the eye. So a given dim flash may contain a small or large number of quanta, and is more likely to be detected if there happen to be more than the average number in the flash. From lengthy experiments to build up 'frequency-of-seeing' curves, the number of quanta can be deduced from the steepness of the curves. So basic physics is important for vision.

The quantal nature of light is important for detecting fine detail. One of the reasons why it is possible to read only the largest newspaper headlines by moonlight is that too few photons fall on the retina to build up a complete image (within the time-span over which the eye can integrate energy, which is about a tenth of a second.) This is pure physics. Sometimes it is hard to establish whether a visual effect should be thought of as belonging to psychology, to physiology, or to the physics of light.

3

Eye

In the beginning

How did eyes evolve? Through what stages did their incredibly sophisticated retinas, their receptors, lenses, irises, and controlling muscles develop? Isn't there a chicken and egg problem here? For what use is an eye without a brain to interpret its images?

The problem of how eyes developed, presented a challenge to the theory of evolution by natural selection which gave Charles Darwin, at the time of the publication of *The origin of species* in 1859, his famous 'cold shudder'. For Darwin's theory of how organisms evolve is very different from inventing and designing with human intelligence. When an engineer sets out to improve an instrument, he can go 'back to the drawing board' and make experimental models, most of which will not work well, if at all; but this is hardly possible for natural selection, for each step must confer some advantage on its owner. What use is a half-made lens? What use is a lens giving an image if there is no brain capable of making effective use of it? How could such a brain come about before there was an eye to feed it information? There is no master plan for evolution; no looking ahead, no experimental inefficient try-outs. Eyes and brains have come about through slow blind trial and error. To retrace the steps we must look for possible advantages at each stage—though allowing that something with one advantage may serve a new quite different use. Who could have guessed that the jaw bone of ancient fish would become the human inner ear? Evolution of the eye and other senses may not be simple linear developments, with each step an improvement on the last for the same function, as new uses may appear. Thus a lens might start as a protective window, to become curved and focus light perhaps millions of years later.

How did eyes start? Almost every living thing is sensitive to light. Plants accept the energy of light, some moving to follow the Sun almost as though flowers were eyes to see it. The first eyes responded only to light and changing intensity of light, with no imaging. Perception of form and colour waited upon more complicated eyes capable of forming images, and brains sufficiently elaborate to interpret their neural signals in terms of objects.

The later, image-forming eyes, developed from light-sensitive spots on the surface of simpler animals. How this occurred is largely mysterious but we do know some of the characters in the story. Some can be seen as fossils; some are inferred from comparative studies of living species; some appear fleetingly during the development of embryo eyes.

Response to light is found even in single-celled organisms. In higher forms we find specially adapted cells to serve as receptors sensitive to movement. These cells may be scattered over the skin (as in the earthworm) or they may be arranged in groups, lining a depression or pit, which is the beginning of a true image-forming eye.

It seems likely that photoreceptors became recessed in pits because there they lay protected from the surrounding glare, which reduced their ability to detect moving shadows heralding approach of danger. The primitive eye pits were open to the risk of becoming blocked by foreign particles lodging within them, so shutting out the light. A transparent membrane developed over the eye pits, serving to protect them. When, by chance mutations, this membrane became thicker in its centre, it became a crude lens. The first lenses served merely to increase intensity: later they came to form useful images. An ancient pit type of eye is still to be seen in the limpet (Figure 3.1). One living creature—*Nautilus*—has an eye still more primitive, for there is no lens but just a pinhole to form the image. The inside of the eye of *Nautilus* is washed by the sea in which it lives; eyes such as ours are filled with a specially manufactured fluid (the aqueous humour) to replace the sea, and human tears are a salty re-creation of primordial oceans which bathed the first eyes.

Here we are concerned with human eyes and how we see the world. Ours are typical vertebrate eyes, and not among the most complex or highly developed, though the human brain is the most elaborate of all brains. Complicated eyes often go with simple brains: in pre-vertebrates we find eyes of incredible complexity serving tiny brains. The compound eyes of arthropods (including insects) consist not of a single lens with a retina of many thousands or millions of receptors; but rather it is many lenses with a small group of receptors, effectively a single receptor for each lens. The earliest known fossil eye belongs to the trilobites, which lived some 500 000 000 years ago.

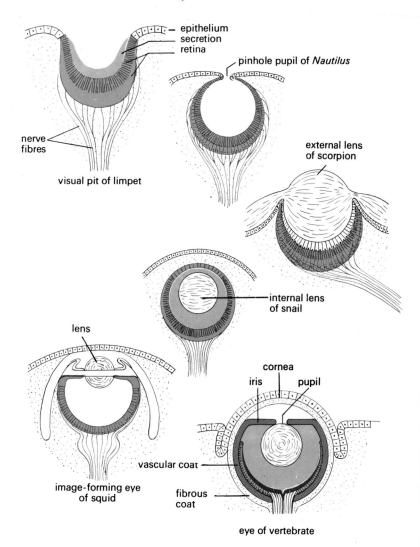

3.1 Examples of primitive eyes. Starting from light-sensitive cells on the skin, a cup forms—becoming a pinhole—then a lens forming an image on a mosaic of tiny tightly spaced light-sensitive receptors.

In many species of trilobites the eyes were highly developed. The external structure of these most ancient eyes may be seen perfectly preserved in fossils (Figure 3.2). Some of the internal structure can be seen in the fossils with X-rays. They were compound eyes, much like those of a modern insect, some with over a thousand facets.

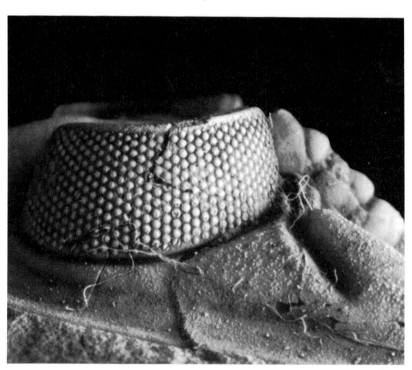

3.2 Fossil compound eye of a 500 million year old trilobite. This is the earliest kind of eye preserved as a fossil. The facets are the corneal lenses, essentially the same as in a modern insect eye. Some trilobites could see all round; but none could see above, as there were no threatening predators in the sea to swoop down on them.

Figure 3.3 shows an insect eye. Behind each lens facet (corneal lens) lies a second lens (lens cylinder) through which light passes to the light-sensitive element, this usually consisting of seven cells grouped in a minute flower-like cluster. Each unit of a compound eye is known as an ommatidium. It used to be thought that each ommatidium is a separate eye—insects seeing thousands of worlds—but it is difficult to understand how this came to be believed, for there is no separate retina in each ommatidium and but a single nerve fibre from each little group of receptors. How, then, could each one signal a complete image? The fact is that each ommatidium signals the presence of light from a direction immediately in front of it, and the combined signals represent what is effectively a single image.

Compound eyes are principally detectors of movement, and can be incredibly efficient as we can see by watching a dragon-fly catching its

ommatidium?

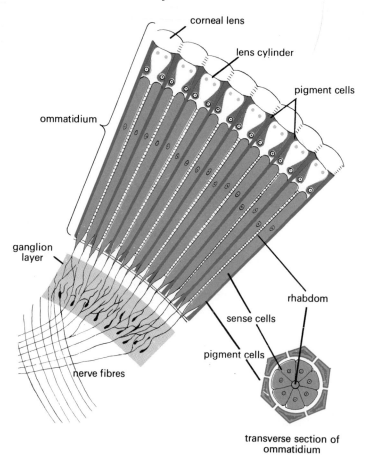

3.3 A compound eye. Each corneal lens provides a separate image to a single functioning receptor (made up often of seven light-sensitive cells), but there is no reason to think that the creature sees a mosaic. The compound eye is especially good at detecting movement, but with low image acuity as they are so small.

prey on the wing. Insect eyes have remarkable mechanisms to adapt to bright or dim light. Ommatidia are isolated from each other by black cones of pigment, which in reduced light migrate back towards the receptors, so that light can pass through the side of each ommatidium to its neighbouring receptors.\This increases the sensitivity of the eye but at a cost to its acuity. This balance is found also in vertebrate eyes, though for somewhat different reasons, as a result of the many neural mechanisms of dark adaption which are still not completely understood.

The lens cylinder of the compound eye does not function simply by virtue of the shape of its optical surfaces, as in a normal lens, but also by change of its refractive index, which increases towards the centre. Light is funnelled through it in a way different from a normal lens, and more like the optics used for medical endoscopes and for channelling laser light over great distances with optical fibres. Nature generally gets there first!

Among the most curious eyes in the whole of nature is that of a creature the size of a pin's head—a little known copepod, *Copilia*. She (the males are dull by comparison) has a pair of image-forming eyes, which seem to function neither like vertebrate nor compound eyes, but like a mechanical television camera. Each eye contains two lenses, and the photoreceptor system is similar to the insect eye, but in *Copilia* there is an enormous distance between the corneal lens and the lens cylinder (Figure 3.4). Most of the eye lies deep within the body, which is extraordinarily transparent.

The scanning movement of the lens cylinder and attached photoreceptor is shown by the successive frames of a cine film in Figure 3.5. The receptors move precisely towards and then away from each other—never independently. The rate of the scan varies from about five per second to one scan every two seconds. The secret of this eye is to be found by looking at the living animal. In 1891, Selig Exner reported that the receptor (and attached lens cylinder) make a 'continuous lively motion'. At that time the concept of sending information down a single channel by scanning was not appreciated—so *Copilia* could not, in the full sense, be seen at that time. Now that we are familiar with how television works we can see such eyes with understanding. It would be fascinating to know whether they are remaining examples of a very early kind of eye. Is *Copilia's* eye ancestral to the compound eye? Is scanning generally abandoned because a single neural link could not transmit sufficient information? Or is it a simplification of the compound eye found in the earliest fossils? If *Copilia* is an evolutionary failure, she deserves a prize for originality; but recent discoveries by Michael Land show that she is not alone, as there are clearer examples of scanning eyes in nature. It is possible that many small compound eyes—such as in *Daphnia*, whose eye has 22 ommatidia and wiggles violently—use scanning to improve the resolution and channel capacity of their few elements, rather like feeling a shape with several moving fingers. However this may be, touch is very important for giving object-meaning to vision.

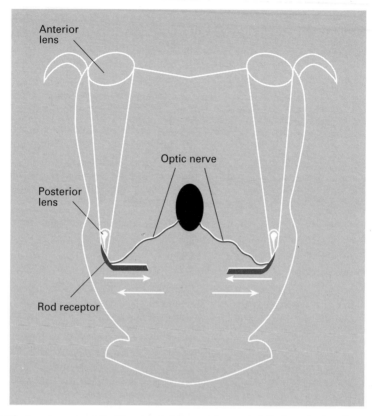

3.4 The female copepod *Copilia quadrata* (above and opposite). Each eye has two lenses—a large anterior lens and a second smaller lens deep in the body with an attached photoreceptor—and single optic nerve to the central brain. The second lens and photoreceptor are in continual movement, across the image plane of the first lens. This seems to be a scanning eye: a mechanical television camera. Over the last few years the biologist Michael Land has discovered several kinds of scanning eyes.

A chicken and egg problem?

What use is an eye without a brain to make sense of its signals? How could such brains develop before there were eyes to provide visual signals? One idea (originating from the author) is that visual processing developed from the immediately useful and essentially simpler processing for touch. It may be suggestive that there are two kinds of touch: moving cilia, or fingers, to explore by *active* ('haptic') touch, and *passive* reception of patterns. Thus one can feel around outlines of things with a finger, or experience the shape of small objects all at once by contact with an area of skin. Could it be that these two kinds

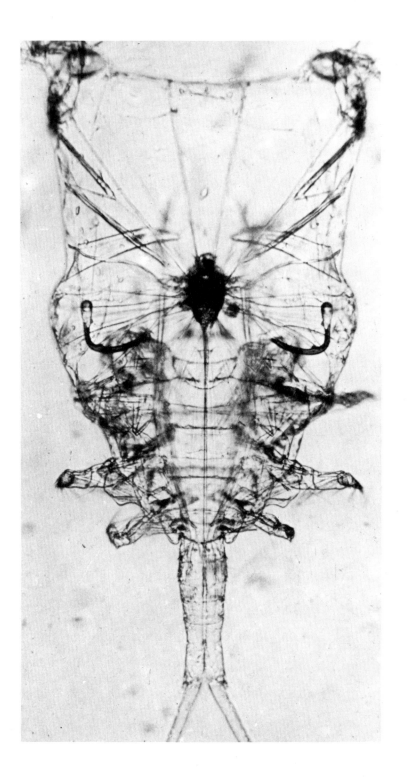

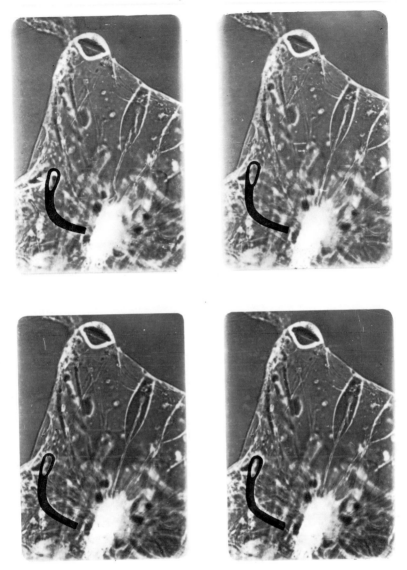

3.5 Frames of a cine film showing scanning of the photoreceptors of *Copilia*.

of touch served for the initial neural processing of the two kinds of eye—compound and simple? Did compound eyes start with a single moving ommatidium, something like *Copilia*? This could convey information at only a slow rate, so perhaps single units multiplied into many ommatidia to give the benefits of parallel processing? It is

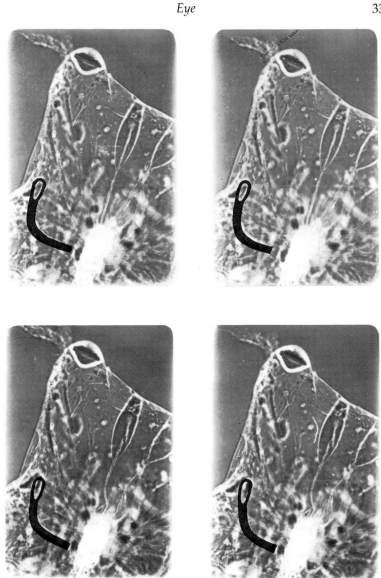

interesting that *Daphnia*, which has 22 ommatidia, is in constant motion; perhaps scanning with its 22 optical fingers. Compound eyes with hundreds or thousands of optical elements do not need to scan, for they have enough units to work in parallel, as simple eyes such as ours do.

Images in eyes

Optical experiments with pinholes, and later with lenses, suggested how the eye works. The Arabian scholar Alhazen (*c.* 965–1038) realized that Euclid (in his *Optics* of about 300 BC) was wrong in thinking of an eye as a geometrical point with rays of light shooting out to touch objects. Alhazen realized that objects can paint themselves on a surface, as an optical image, with just a pinhole and a screen (Figure 2.1). For, in the tenth century, Alhazen devised the first *camera obscura* (darkened room). This was developed by Giovanni Battista della Porta, described in his *Natural magic* of 1558, by replacing the pinhole with a focusing lens to give images sufficiently bright to see form and colour clearly (Figure 3.6).

He suggested that this is how the eye works:

The image is let in by the pupil, as by the hole of a window; and that part of the Sphere, that is set in the middle of the eye, stands instead of a crystal Table. I know ingenious people will be much delighted by this.

The effectiveness of the camera obscura, which revealed the optics of the eye, was well described by Damielo Barbaro in his *Practica della perspectiva* (1568–69):

Close all the shutters and doors until no light enters the camera [room] except through the lens, and opposite hold a sheet of paper, which you move for-

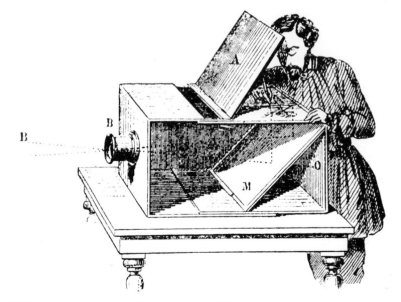

3.6 The camera obscura—producing an image with a lens.

ward and backward until the scene appears in the sharpest detail. There on the paper you will see the whole as it really is, with its distances, its colours and shadows and motion, the clouds, the water twinkling, the birds flying. By holding the paper steady you can trace the whole perspective with a pen, shade it and delicately colour it from nature.

The secret was out. The world is seen by the brain from images in the camera obscura of the eye. This was appreciated by the astronomer Johannes Kepler and the philosopher–mathematician René Descartes (Figure 3.7).

So at last eyes could be thought of as optical devices—mechanisms obeying laws of physics and optics. It was an important step philosophically as well as being practically useful, revealing the nature of images and perspective for pictures.

The perfection of the eye as an optical instrument is a token of the importance of vision in the struggle for survival. Not only are the parts of the eye beautifully contrived, but the tissues are specialized, and they are nurtured and protected in special ways (Figure 3.8). The

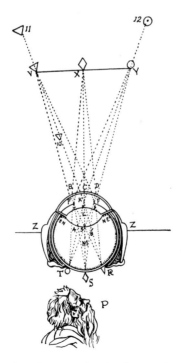

3.7 René Descartes's drawing of an eye showing how an image is formed. From *La dioptique* (1637).

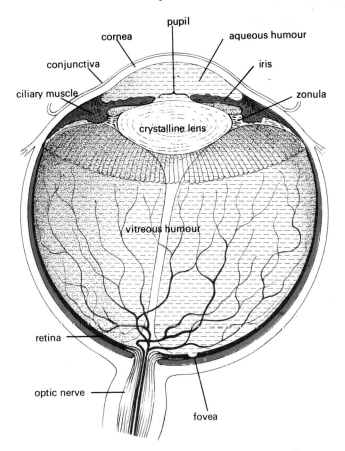

3.8 The human eye. The most important optical instrument. Here lies the focusing lens, giving a minute reversed image on a dense mosaic of light-sensitive receptors, which convert the patterns of light energy into the language the brain can read—chains of electrical impulses.

cornea is unusual in having no blood supply, obtaining its nutrient from the aqueous humour instead. Because of this the cornea is virtually isolated from the rest of the body—a factor that makes corneal transplants particularly safe since antibodies do not reach and destroy the transplanted tissue. The aqueous humour is continually secreted and absorbed, so that it is renewed about once every four hours. 'Spots before the eyes' can be due to floating impurities casting shadows on the retina, which are seen as hovering in space.

Another specialized structure isolated from the blood stream is the lens. A structure in the inner ear is similarly isolated, although the significance of this is different. In the cochlea, where vibrations are converted into neural activity signalling sound, the remarkable structure known as the organ of Corti—rows of very fine hairs joined to nerve cells that are stimulated by the vibration of the hairs—receives its nutrients from a fluid filling the cochlea. If these very sensitive cells were not isolated from the pulse of the arteries, we would be deafened.

Each eyeball is equipped with six extrinsic muscles, which hold it in position in its orbit and rotate it to follow moving objects, or direct the gaze to chosen features. The eyes work together so that normally they are directed to the same object, converging for near objects and parallel for great distance. There are also muscles within the eyeball. The iris is an annular muscle forming the pupil, through which light passes to the lens, which lies immediately behind. The iris contracts to reduce the optical aperture of the lens in bright light, and also when the eyes converge to view near objects. This increases the depth-range of sharp focus, in the same way as 'stopping down' a camera lens. Another muscle controls the accommodation (or focusing) of the lens. We will look in more detail at the mechanism and function of the lens and the iris. Both have surprises.

The crystalline lens

It is often thought that the lens serves to bend the incoming rays of light to form the image. This is far from the whole truth in the case of the human eye, though it is so for fishes. The region where light is bent most in the human eye, to form the image, is not the lens but the front surface of the cornea. The reason for this is that the power of a lens to bend light depends on the difference between the refractive index of the surrounding medium and the lens material. The refractive index of the air is very low, while that of the aqueous humour immediately behind the cornea is nearly as high as the material of the lens. So the cornea is primarily important for focusing, the lens serving to adjust for different distances. The case of fish is different because the cornea is immersed in water, so light is hardly bent when it enters the eye. Fish have a very dense rigid lens which is spherical, moving backwards and forwards within the eyeball to focus distant or near objects. In human and other land animals' eyes, accommodation to distance is done not by changing the position of the lens (as in a fish, or a camera), but by changing its shape. The radius of curvature is reduced for near vision, making the lens increasingly powerful and so adding more to the primary bending of light at the cornea.

There is an eye—of a fish named Anableps—which has a lens that

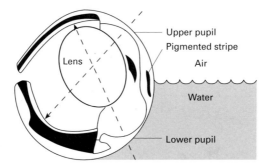

3.9 Anableps—an eye which focuses in air and water.

focuses simultaneously for air above and water below (Figure 3.9). The lens is elliptical, arranged so that light from the air above passes through a smaller radius of curvature than the light from the water below. The bodies of Anableps are asymmetrical: the 'right-handed' cannot mate with the 'left-handed'; but both have these extraordinary eyes, having two powers of accommodation.

The human lens is built up of thin layers, like an onion, and is suspended by a membrane, the zonula, which holds it under tension. Accommodation works in a most curious manner. For near vision, tension is reduced on the zonula by contraction of the ciliary muscle, allowing the lens to spring to a more convex form. Thus the lens accommodates for near vision by the muscle tightening, not relaxing, which is a surprising system. Possibly this 'backwards' mechanism avoids muscle tremor on the lens.

The embryological and later developments of the lens are of particular interest, and have dire consequences in middle life. The lens is built up from its centre, cells being added all through life, so it continues to get larger. The centre is the oldest part, and there the cells become increasingly separated from sources of oxygen and nutrient, and eventually die. When they die they harden; so after the age of about 40, most of us have to start using spectacles as the lens becomes too stiff to change its shape for accommodation to different distances. This is all too clear in Figure 3.10, which shows how accommodation falls off with age, as the starved cells inside the lens die and we see through their corpses.

It is possible to see the changes in shape of another person's lens as they accommodate to different distances. This requires no apparatus beyond a small source of light, such as a flashlight. If the light is held in a suitable position, it can be seen reflected from the surface of the

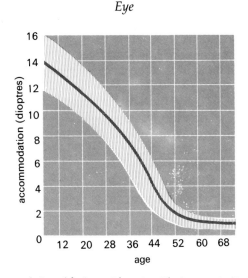

3.10 Loss of accommodation of the lens with ageing. The lens gradually becomes rigid, so cannot change its shape.

eye; but there is not just one reflection—there are three. The light is reflected not only from the cornea but also from the front and the back surfaces of the lens. As the lens changes its shape, these images change in size. The front surface gives a large and rather dim image, which is the right way up, while the back surface gives a small, bright, upside-down image. The principle can be demonstrated with a spoon (see box).

Spoon reflections in the eye

Reflected from the back convex surface you will see large right-way-up images, but the inside concave surface gives small upside-down images. The size of the images will be different for a large (table) spoon and a small (tea) spoon—corresponding to the curvatures of the lens of the eye for distant and near vision. These images seen in the eye are known as Purkinje images, and are useful for studying accommodation experimentally.

The iris

This is pigmented, and is found in a wide range of colours. Its Greek meaning—rainbow—is poetically appropriate. The colour of a person's eyes is of interest to geneticists and lovers, but less to those concerned with the function of the eye. It hardly matters what colour the iris is, but it must be reasonably opaque so that it is an effective aperture stop for the lens. Eyes where pigment is missing (in albinism) are, for this and other reasons, seriously defective.

It is sometimes thought that the changes in pupil size are important in allowing the eye to work efficiently over a wide range of light intensities. This could hardly be its primary function, however, for its area changes over a ratio of only about 16:1, while the eye works efficiently over a range of intensity greater than 1 000 000:1. It seems that the pupil contracts to limit the rays of light to the central and optically best part of the lens, except when the full aperture is needed for maximum sensitivity. As we have said, it also closes for near vision, increasing the otherwise limited depth of field for near objects, just as in a camera.

To an engineer, any system which corrects for an external change (in this case light intensity) suggests a servomechanism. These are very familiar, for example in the form of the thermostats controlling central heating, which switch on the heat automatically when the temperature drops below some pre-set value and then switch it off again when the temperature rises. An early example of a man-made servomechanism is the windmill, which aims into the wind and follows its changing direction by means of a fantail sail which rotates the top of the mill through gearing. A more elaborate example is the autopilot which keeps a plane on correct course and height by sensing errors and sending correcting signals to the control surfaces. This is crude, however, compared with bird flight.

To go back to the thermostat sensing temperature changes in a central heating system: imagine that the setting of the lower temperature for switching on the heat is very close to the upper temperature for switching it off. No sooner is it switched on, than the temperature will rise enough to switch it off again—so the heating system will be switched on and off rapidly, until perhaps something breaks. By noting how frequently it is switched on and off, and noting also the amplitude of the temperature variation, an engineer could deduce a great deal about the system. With this in mind some subtle experiments have been performed to reveal how the iris servocontrol system of the eye works.

The iris can be made to go into violent oscillation by directing a narrow beam of light into the eye so that it passes the edge of the iris (Figure 3.11). When the iris closes a little the beam is partly cut off and the retina gets less light, so the iris starts to close until it gets another signal to open—oscillating in and out. By measuring the frequency and amplitude of oscillation of the iris, a good deal can be learned about the neural servosystem controlling it.

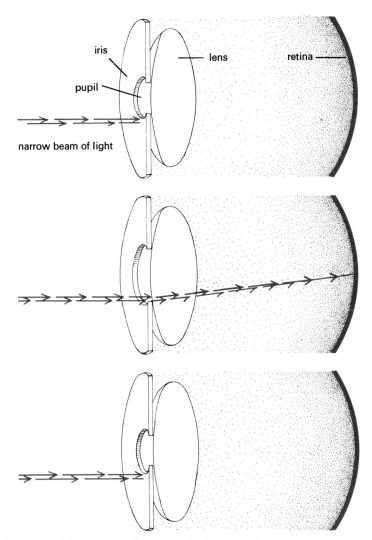

3.11 Making the iris oscillate with a beam of light. When the iris opens a little, more light reaches the retina, which then signals the iris to close. But when it starts to close, less light reaches the retina, which signals the iris to open. Thus it oscillates. From the frequency and amplitude of oscillation, the iris control system can be described in terms of engineering servo-theory.

The pupil

The central hole of the iris—the pupil—looks black and normally we cannot see through it into another person's eye. This requires some explanation, for the retina is not black, but pink. Indeed it is curious that although we see out of our pupils we cannot see into someone else's! The reason is that, as one's own eye is focused on the other's, one's eye and head get in the way of the light, just where it is needed to see the other's retina (Figure 3.12).

The ophthalmoscope, for looking into eyes, works by directing light into the other eye (Figure 3.13). With this device the pupil no longer looks black, but pink, and the detailed structure of the living retina can be seen. The blood vessels on its surface appear as a great red tree, with many branches (Figure 3.14). Potentially sinister changes or abnormalities may be seen in time for prevention or cure. In fact the retina is the only living structure of the body that can be seen non-invasively, as the surface of the skin is not alive, so the ophthalmo-scope is a generally useful diagnostic tool for medicine, as the state of blood vessels can be seen. It was invented by Helmholtz in 1850.

The annoying 'red-eye' of flash photography occurs when the subjects' eyes are not precisely accommodated to the distance of the camera, so that unfocused light from the retina reaches the camera.

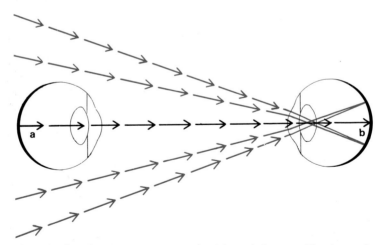

3.12 The eye at the left cannot see into eye at the right, and vice versa. The observer's eye (and head) gets in the way, preventing light reaching the region of the other retina where it should be imaged by the observer's eye. Neither eye can see the other's retina, for each hides the light. Similarly, one can't see into a camera through its lens, when it is focused at one's eye.

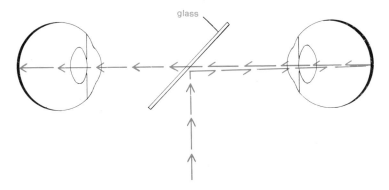

glass

3.13 The ophthalmoscope. Light reaches the retina by reflection from a part-silvered mirror, through which the observer sees inside the now illuminated eye. In practice the observer may look just above the illuminating ray, to avoid glare from the mirror which can be fully silvered. Lenses in the instrument give sharp focus.

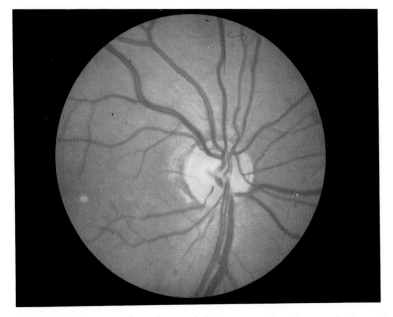

3.14 A normal retina seen through an ophthalmoscope. This photograph shows the yellow spot over the fovea, where acuity is highest, the retinal blood vessels through which we see the world, and the large blind spot where the blood vessels and nerves leave the eyeball.

This is the basis of a useful technique for seeing whether infants' eyes have normal accommodation.

Blinking

Blinking is often assumed to be a reflex, initiated by the cornea becoming dry. It is true that blinking can be initiated by irritation of the cornea, or by sudden changes in illumination, but normal blinking occurs with no external stimulus: it is initiated by signals from the brain. The frequency of blinking increases under stress, and increases with expectation of a difficult task. It falls below average, however, during periods of concentrated mental activity. Blink rates can be used as an index of attention and concentration on tasks of various kinds and difficulties. A hazard of constant-concentration tasks, such as engineering or architectural drawing, is that the corneas can dry up through reduced blinking, with a long-term risk of damage to the corneas of eyes.

Eye movements

Each eye is rotated by six muscles (Figure 3.15). The remarkable arrangement of the superior oblique can be seen in the illustration. The tendon passes through a 'pulley' in the skull, in front of the suspension of the eyeball. The eyes are in continuous movement, and they move in various ways and generally not smoothly. When the eyes are moved around searching for an object they move quite differently from when a moving object is followed. When searching they move in a series of small rapid jerks: when following, they move smoothly. The jerks are known as *saccades* (after an old French word meaning 'flick of a sail'). Apart from these two main types of movement, there is also continuous small high-frequency tremor.

Movements of the eyes can be recorded in many ways (see box).

As high acuity in humans (though not all species) is restricted to the central foveal region, eye movements are needed for selecting what to look at clearly; so recording eye movements tells us what the brain thinks it needs in any situation. The movements are affected by what we are doing and by what we are looking at. A Russian psychologist, Alfred Yarbus, is the pioneer in recording movements of the eyes while looking at pictures (Figure 3.16). The eyes move in saccades favouring certain features of a picture or scene. Records of what is selected tell us a lot about how perceptions are built up from successive fixations.

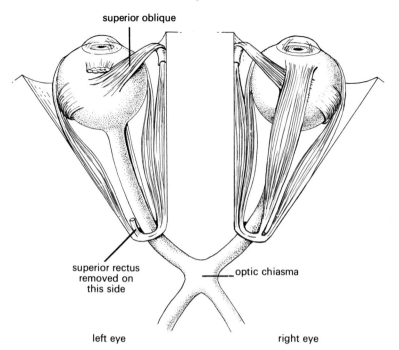

superior oblique

superior rectus removed on this side

optic chiasma

left eye   right eye

3.15 Muscles moving the eyes. The eyeball is maintained in position in the orbit by six muscles, which direct the gaze to any position, and allow the eyes to converge at varying distances, from a few centimetres to infinity. The muscles are under constant tension, forming a delicately balanced system, which when upset gives disturbing illusions of movement.

Recording eye movements

Eye movements can be recorded:

(1) with a cine or a TV camera;

(2) by surface electrodes on the skin near the eyes (as the eyeballs have a small standing potential between retina and cornea);

(3) by a coil attached to a contact lens (small voltage changes can be detected);

(4) by attaching a tiny mirror to a contact lens on the cornea, when a beam of light may be reflected off the mirror, and photographed on continuously moving film (this is the most accurate method, but can be dangerous).

There are also optical systems for receiving reflected infra-red light from the eyes, for recording their movements or stabilizing selected parts of images with a servosystem, and other sophisticated methods, though none is ideal.

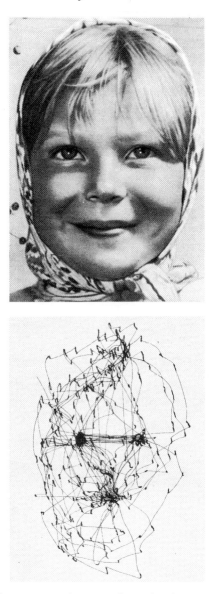

3.16 How eyes search a picture. Such records show what the brain thinks it needs. This depends on what it is seeing and the current task. Movements when driving a car are very different from reading, or looking at pictures such as this. For portraits, eyes and noses are specially selected. (From Yarbus 1967.)

There is evidence that signals from the eyes are inhibited during the rapid saccades of eye movements. Presumably this prevents the blur from such rapid movement degenerating retinal signals, and may also contribute to why the world does not seem to spin round during movements of the eyes (pages 101–109).

Seeing the blindness of saccades

Look at your eyes in a mirror, looking from one eye to the other. It is impossible to see them move. Yet one person can see the movements of another's eyes. Ones own saccades suppress signals from the retinas.

You will also find that any small after-image in the fovea will disappear during each saccadic eye movement.

A surprising phenomenon has been found recently, during research into eye movements while reading and looking at pictures. James Grimes presented subjects with alternating pairs of pictures which were largely the same, but had some differences. He found that, when a saccade occurred between alternated pictures, even large differences were not noticed. For example if people's hats, or heads, were switched (now easy to do with computer paint programs), or brilliant colours of bathing dresses, or birds were changed, subjects took many seconds to see a difference (Figure 3.17). If pictures were alternated following a saccade, differences were unnoticed if they affected features which were not receiving particular attention. This stresses the importance of internal representations for perception; far less is registered at each fixation of the eyes than we realize.

Fixing retinal images

It is possible to stabilize an image on the retina, so that wherever the eye moves, the image moves with it. When optically stabilized, vision soon fades (Figures 3.18). An important function of eye movements is to sweep the image over the receptors so that they do not adapt and cease to send signals to the brain.

Now we meet a curious problem. When we look at a blank sheet of paper, the edges of its image will move around on the retina, so stimulation of these receptors will be renewed. But consider the centre of the image: here movements of the eyes can have no effect, for a region of given brightness is substituted for another region of exactly

3.17 The same? If a saccadic eye movement occurs between pictures, it can take several seconds to see what has happened.

the same brightness, so there will be no change of stimulation. Yet the paper does *not* fade away. Why this is so, is not fully understood, but it suggests that it is primarily borders that are signalled. Areas of constant intensity seem to be inferred from surrounding borders. This is an economical way of signalling images.

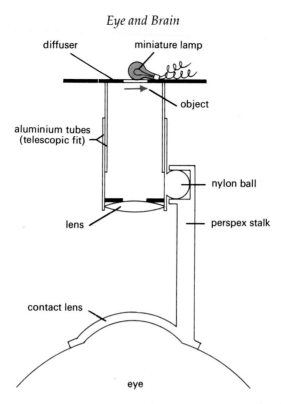

3.18 A simple—but dangerous—way of stabilizing retinal images. The object, a small photographic transparency, is carried on a corneal lens moving exactly with the eye. The eye soon becomes blind to a stabilised image. Some parts generally fade before others. (This heroic method was devised by Roy Pritchard.) Selective fading can be seen in after-images.

Making spectacles for ourselves

The great invention of eyeglasses (Figure 3.19) was not based on the science of the time, but from craftsmen playing about with lenses, probably Venetian glass workers in about 1286. This made spectacles socially suspect to the aristocracy for hundreds of years, and unfortunately scholars associated weak eyes with weak brains, so they were worn in secret. The earliest spectacles were convex, for short sight. Concave lenses were not available before the middle of the fifteenth century.

Eyeglasses are perhaps the invention, next only to fire, that brings most aid and comfort. And what else works forever with no maintenance and no energy costs? Spectacles lengthen our effective lives as with their aid we can see to read and to perform skilled tasks into old

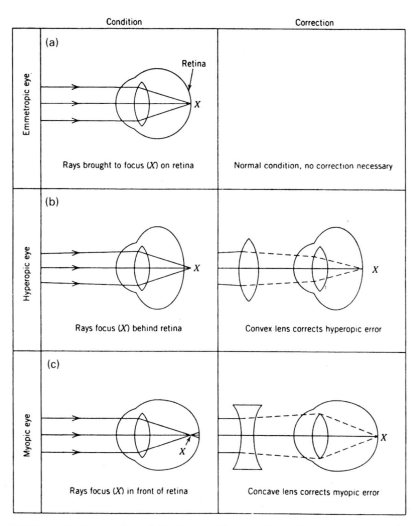

Condition	Correction
(a) Emmetropic eye	
Rays brought to focus (X) on retina	Normal condition, no correction necessary
(b) Hyperopic eye	
Rays focus (X) behind retina	Convex lens corrects hyperopic error
(c) Myopic eye	
Rays focus (X) in front of retina	Concave lens corrects myopic error

3.19 Correcting long or short sight. Glasses add to, or subtract from, the focusing of the cornea and to a lesser extent the lens of the eye.

age. Before they were available, scholars and craftsmen were made helpless by lack of sight just as their skills matured.

The discovery that the lens of the eye works by simple physics—exactly like a glass lens—opened the mind to seeing our bodies as understandable through experiments and by analogies with the physical world. Seeing the eyes' lenses as within physics, was a significant step away from the vitalism which blocked biological understanding, holding that every aspect of life is essentially unique and so beyond explanation. Our uniqueness is surely being able to question and discover, and sometimes to explain ourselves and what we see.

Focusing on the retina

The term 'retina' is derived from an early word meaning 'net', or ' cobweb tunic', from the appearance of its blood vessels. It is a thin sheet of interconnected nerve cells, including the light-sensitive rod and cone cells which convert light into electrical pulses—the language of the nervous system. It was not always obvious that the retina is the first stage of visual sensation. The Greeks thought of the retina as providing nutrient to the jelly filling the eye, the vitreous humour. The source of sensation was supposed by Galen (*c.* AD 130–201), and by much later writers, to be the crystalline lens. The Arabs of the dark ages—who did much to preserve and develop Greek notions of optics—thought of the retina as conducting the supposed vital spirit, the 'pneuma'.

It was the astronomer Johannes Kepler who, in 1604, first realized the true function of the retina: that it is the sensitive screen on which images from the lens are projected. It is believed that this hypothesis was tested experimentally by C. Scheiner in 1625, by cutting away the outer coating (the sclera and the choroid) from the back of an ox's eye, leaving the retina revealed as a semi-transparent film, with a reversed image on the retina in the ox's eye. Descartes tried it, and described it in his *La dioptique* of 1637 with clear diagrams (Figure 3.7). It is remarkable how late in the history of science this discovery came, and perhaps even more remarkable how long it has taken to appreciate the full richness of its implications. Even now there are experts on vision who try to maintain that perception is directly in contact with the world of objects—in spite of the optical screen of the retina, and the vast physiological complexity of image processing (with knowledge in mind) creating perceptions. To be so isolated from the world of objects can be quite frightening. And as we will see, it opens the way to all manner of illusions.

For many years, the inversion of the retinal image seemed to be a serious problem: do babies have to learn to correct the inversion, to see things the right way up? This question is based on the false notion that the brain is a kind of eye looking at the retinal image, which presumably has its image—with another eye, another image, and so on for ever. This notion of an inner eye should have been a non-starter as it can never get anywhere. The point is, one's retinal image is not seen, as an object is seen. The retina is the interface between the optical projection from objects to the neural-coded signals to the brain—arriving down the million fibres of the optic nerve—which are related to touch experience of objects. The inversion in the image does not matter: what matters is the relation of the brain's visual signals to those from touch. Indeed without touch retinal images would have little or no meaning. The image, however, is represented as a rough map at the first stage of visual processing, and it is suggestive that the corresponding touch map is also upside-down (so the head is represented below the feet) presumably to keep the connections between vision and touch as short as possible (see Figure 4.2).

Another confusing problem was the important fact that the sizes of retinal images are given simply by the size and distance of objects, by the subtended angle—yet distant objects generally appear larger than this would suggest. Also, why should objects appear at particular distances and specific sizes, although there is an infinite combination of sizes and distances giving the same angle and the same size of retinal image? That the images have this ambiguity, seldom revealed in perception, puzzled Kepler and is still not completely understood.

The discovery of photoreceptors had to wait upon the development of the microscope. It was not until about 1835 that they were first described, and then none too accurately, by G. R. Treviranus. It seems that his observation was biased by what he expected to see, for he reported that the photoreceptors face the light. Strangely, they do not: in mammals—though not in cephalopods—the receptors are at the back of the retina, behind the blood vessels. This means that light has to go through the web of blood vessels and the fine network of nerve fibres—including three layers of cell bodies and a host of supporting cells—before it reaches the receptors. Optically, the retina is inside out, like a camera film put in the wrong way round (Figure 3.20). Given the original 'mistake' however (which seems to result from the embryological development of the vertebrate retina from the surface of the brain), the situation is largely saved by the nerve fibres from the periphery of the retina skirting around and avoiding the crucial central foveal region giving best vision (see below).

The retina has been described as an outgrowth of the brain. It is a specialized part of the surface of the brain which has budded out and become sensitive to light. It retains typical brain cells which are functionally between the receptors and the optic nerve (but situated in the front layers of the retina), and greatly modify the electrical

LIGHT

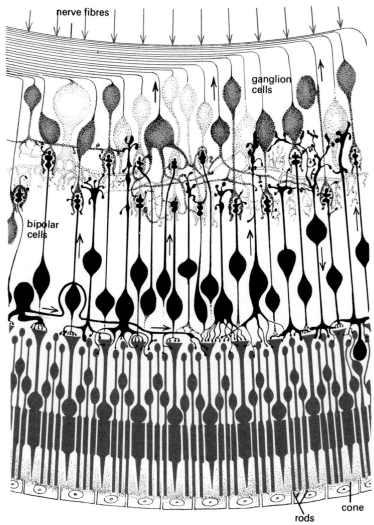

3.20 Structure of the retina. Light travels through the layers of blood vessels, nerve fibres and supporting cells to the sensitive receptors (rods and cones). These lie at the back of the retina, which is thus functionally inside-out. The optic nerve is not, in vertebrate eyes, joined directly to the receptors, but is connected via layers of highly interconnected cells (ganglion, bipolar), which are part of the brain living in the eyes.

activity from the receptors themselves. Some of the data processing for perception takes place in the eye, which is thus an integral part of the brain. This pre-processing funnels 120 000 000 receptors down to 1 000 000 optic nerve fibres, no doubt reducing the thickness and stiffness of the optic nerve so that eye movements are possible.

There are two kinds of light-receptor cells—the *rods* and the *cones*—named after their appearance as viewed with a microscope. In the peripheral regions of the retina they are clearly distinguishable; but in the central region, the fovea, the receptors are packed exceedingly close together, and all look like rods.

The cones function in daylight conditions, and give colour vision. The rods function under low illumination, giving vision only of shades of grey. Daylight vision, using the cones of the retina, is referred to as *photopic* while the grey world given by the rods in dim light is called *scotopic*. Between the brightness of sunlight and the dim light of the stars is the intermediate light from the moon, giving uneasy *mesopic* vision, which should not be trusted.

The receptors are cross-connected, to give 'receptive fields', which are larger in dim light. This sacrifices acuity of fine detail for increased sensitivity, rather like putting a faster but more grainy film in a camera for photographing in dim light. In the eye, however, this is a neural mechanism. Another, *lateral inhibition*, might be thought of as neural sharpening of images, as strongly stimulated receptors inhibit their less stimulated neighbours; but it is better to think of this mechanism as reducing unimportant signals to the brain, as edges are highly significant. The eye even creates edges that are not there, with a phenomenon known as Mach's bands which are seen on a textureless brightness gradient.

How do we know that only the cones mediate colour vision? This is deduced partly from studies of various animal eyes, by relating retinal structure to their ability to discriminate colours as determined by behavioural experiments, and also from the finding that in the human retina there are very few cones near the edge of the retina, where there is no colour vision. It is interesting that although the central foveal region, which is packed tightly with functional cones, gives the best visual detail and colour it is less sensitive than the more primitive rod-regions of the retina. (Astronomers 'look off' the fovea when they wish to detect very faint stars, so that the image falls on a region of the retina rich in sensitive rods. But this does not reveal the colours of the stars.)

It might be said that whenever we look from the central fovea towards the periphery we travel back in evolutionary time—from the most highly organized structure in nature to a primitive eye barely capable of detecting movements of shadows. The very edge of the human retina does not even give a sensation when stimulated by movement. It provides unconscious reflexes, directing the highly developed foveal region to where it is likely to be needed, to use its high acuity and massive brain power for establishing what might be there. In this everyday sweep from periphery to fovea, our eyes replay the most dramatic journey through hundreds of millions of years of biological time.

The minute size of the receptors and their extraordinary packing density become significant when we consider the ability of the eye to distinguish fine detail. We shall quote directly from the Spanish histologist S. L. Polyak's *The retina* (1941):

The central territory where the cones are almost uniformly thick measures approximately 100μ across, corresponding to 20′, or one-third of a degree of arc. It contains approximately fifty cones in a line. This area seems to be not exactly circular but elliptical, with the long axis horizontal, and may contain altogether 2,000 cones . . . the size of each of the 2,000 receptor-conductor units measures, on the average, 24″ of arc. The size of the units even in this territory varies, however, the central most measuring scarcely more than 20″ of arc or even less. Of these—the most reduced cones, and therefore the smallest functional receptor units—there are only a few, perhaps not more than one or two dozen. The size of the units given includes the intervening insulating sheaths separating the adjoining cones from one another.

It is worth trying to imagine the size of the receptors. The smallest is 1 μm, only about two wavelengths of red light in size. One could not ask for much better than that. Even so, the image of the smallest separation of lines that can be distinguished is many times less than the diameter of a single cone cell. The brain uses differences of intensity signalled from many receptors, so acuity for lines is better than for points. The acuity of the hawk is four times better than ours; but its fovea lies on the walls of a deep pit, so the image spreads across more receptors, though possibly at a cost of distortions. In general, the human eye is multipurpose, lacking many of the specializations of other eyes, so visual information is relatively unselected for special needs, allowing our enormous brains to select from a wide range of relatively uncontaminated data.

The number of cone receptors is about the same as the population of Greater New York. If the whole population of the United States of America were made to stand on a postage stamp, they would represent the rods on a single retina. As for the cells of the brain—if people were scaled down to their size we could hold the population of the earth in our cupped hands, but there would not be enough people on earth to make one brain.

Seeing inside our own eyes

After-images—photographs in the eye

The photopigments of the retina are bleached by bright light; it is this bleaching which, by some still mysterious process, stimulates the receptors. It takes some time for the photochemicals to return to normal—up to almost an hour for the rhodopsin of the rods and about seven minutes for the iodopsin of the cones. While a region is bleached, this region of the retina is less sensitive than the surrounding regions, giving negative *after-images*. Thus when the eye has been adapted to a bright light (for example, a lamp bulb viewed with the eye held steady or, better, a photographic flash) a dark shape of the same form as the adapting light is seen hovering in space. It is dark when seen against a lighted surface, but for the first few seconds it will look bright, especially when viewed in darkness. This is called a positive after-image, and represents continuing firing of the retina and optic nerve after the stimulation. When dark, it is called a negative after-image, and represents the relatively reduced sensitivity of the previously stimulated part of the retina, due partly to bleaching of the photopigment. Exactly what happens is extremely complicated and the processes giving increased sensitivity in dim light—dark adaptation—are not fully understood. They are not (as might be expected) simply related to the loss of photochemical molecules with bleaching, as this is insufficient to explain the enormous range of adaptation (something like a million to one) that occurs. There must also be a neural gain control, somehow set by the amount of bleaching; but just how this works is still a puzzle.

It is easy to see one's own retinal blood vessels and nerve fibres (see box). As the retinal nerve cells are really part of the brain in the eyes, here we are seeing parts of our own brains!

Not all eyes are 'inside-out'. The retinas of octopus have their receptors in front of the blood vessels and nerve cells and fibres. The reason for the poor design of vertebrate eyes is embryological: as the eye buds off from the brain it was simply too hard to reverse the structure. So looking at our retinas we see a failure of evolution.

Each retina has a surprisingly large blind region where the nerves

---

I Iow to see inside your own eye

Place a small torch (or flashlight) just touching the outer corner of one closed eye, in a dark room. Then waggle the light gently from side to side. A great tree of retinal blood vessels will become visible.

This happens because the light shifts across the structures above the receptors, producing moving shadows. Normally the shadows of the blood vessels are stable because the light enters at fixed angles through the lens. The receptors very rapidly adapt to the small differences of intensity, so the overlying structures are not signalled. The moving light, penetrating the eye beyond the lens, beats the system to reveal normally invisible structures of the eye to its owner.

Retinal nerve fibres can be seen by looking through a small hole in an opaque piece of paper, held close to the eye. If you look at a well-illuminated white wall, and jiggle the paper so that the hole crosses the pupil, a fine structure of retinal nerve fibres is seen. This is because the moving pinhole allows light of changing angles through the lens to the retina. When the jiggling is stopped the structure fades away quite slowly. This shows the time-constant of adaptation of the receptors, as they cease to signal the small differences of opacity of the almost transparent fibres above them.

---

and blood vessels leave the eyes (Figure 3.14), called the blind spots. They are so large one can make a person's head disappear, when they are sitting across the room! The effect is most easily seen (or rather not seen) with a pair of dots, or marks such as these:

$$\bigcirc \qquad\qquad\qquad\qquad +$$

Close your left eye, look straight at the $\bigcirc$, and move your head slowly backwards and forwards, until (at about 20–30 cm) the + will disappear. Then try it with the left eye open, fixating the +. The $\bigcirc$ disappears when the image falls on the blind spot, which has no light-sensitive receptors.

Here there is a puzzle: why is this page (or any surface) seen in this blind region of the eye? The other eye can't be helping, for it is closed.

Why don't we see scotomas?

There are two possibilities. The brain might be *ignoring* the blind region, as it never gets signals from here. Or, the brain might be *constructing* what 'ought' be there, on the evidence of the surrounding colour and pattern. It is interesting to repeat the experiment with a

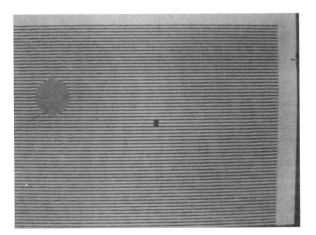

3.21 Experiment to demonstrate filling-in blind regions. When the eyes are aimed at the 'fixation point' of the first display, the blind region fades, disappearing in about 10 seconds. This patterned display is switched off, and the blank screen switched on. A region of the pattern is now seen in the place of the blank region of the first display. This seems to be the brain's creation for filling in the blank region, or scotoma. (From Ramachandran and Gregory 1991.)

complex pattern to investigate how complicated it can be for filling-in to be possible. This is not entirely clear cut as visual acuity is not high around the peripheral regions of the blind spots.

If, instead, the blind spots are not filled in but simply ignored, this might be because they have never provided information. But what happens with a new scotoma?

It is possible to make an artificial blind region or scotoma at any time, safely, in any region of the retina. This can be done simply by staring at a 'fixation spot' placed at some distance from a smallish blank region in a pattern. The blank region gradually disappears, becoming invisible after about 10 seconds. Is it now ignored, or actively filled in from the surrounding pattern?

Filling in is confirmed by a recent experiment, using computer graphics which allows a pattern to be rapidly replaced by a blank screen of similar brightness and colour. Subjects view a patterned display with a blank region (Figure 3.21). When the blank region has disappeared, the display is replaced by the blank screen—which has nothing except the fixation spot to prevent the eyes moving. The result is remarkable. A small bit of the switched-off pattern is seen on the blank screen, in the region of the missing bit of pattern of the first display, though somewhat degenerated. The same happens with a dynamic moving pattern (twinkly visual 'noise') instead of the stationary fixed pattern. Then a patch of twinkle is seen on the blank screen, in the place of the previous missing region.

This effect is entirely different from familiar after-images (page 57). We seem to be seeing our brain's creation of the pattern, normally serving to fill scotomas, so that we are not distracted by fictional black blobs from the blind spots in the field of vision. This process occurs quite early, for it takes no account of object-knowledge. Thus a missing nose is not added: The process works merely for patterns, not objects. It can function early on in visual processing, because in any circumstances it is useful to avoid visible scotomas—which would look like threatening black objects—freeing later processing from having to edit them out.

Perceiving depth from two eyes

Many of the organs of the body are duplicated but the eyes and also the ears are unusual in working in close co-operation. They share and compare information, and together they perform feats impossible for a single eye or ear—especially signalling distance or depth for vision and direction for hearing. The images in the eyes lie on the curved surfaces of the retinas but it is not misleading to call them two-dimensional. A remarkable thing about the visual system is its ability to combine the two somewhat different images, into a single perception of solid objects lying in three-dimensional space (stereoscopic perception).

For eyes such as ours, with a concentration of receptors in central vision, the eyes are aimed to bring images of near and far objects onto the foveas. This vergence of the eyes gives best vision for each eye and also, from the angle of convergence, signals distance like a simple range-finder (Figure 3.22).

Depth is signalled by the angle of convergence, but there is a serious limitation to any such range-finder. It can indicate distance for only one object at a time. To signal distances of many objects simultaneously, a far more complicated system is needed, because different disparities of many objects must be signalled, independently of the vergence angle of the eyes.

Stereo (3-D) depth

The eyes are horizontally separated by about 6.5 cm, so receive different views for near objects. This can be seen clearly if each eye is closed alternately. Any near object will appear to shift sideways in relation to

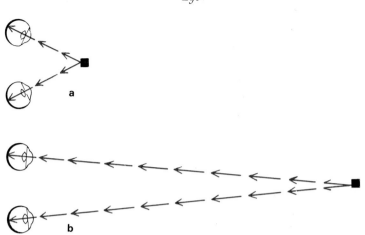

3.22 Range-finder convergence. The images are centralized on the foveas for different distances: (**a**) converged to a near object; (**b**) to greater distance. The angle of convergence provides information of distance, as for a simple range-finder.

more distant objects and to rotate slightly when each eye receives its view. This always quite small difference between the images is known as disparity. It gives perception of depth as stereoscopic vision. This is employed in the stereoscope, which reveals much of how it works. It is surprising how recently stereoscopic vision has been recognized. Stereo vision only functions for quite near objects, because, with increasing distance, differences between the images become too small. We are effectively one-eyed for objects further than about 100 metres.

The stereoscope was invented just before photography, by the English physicist Sir Charles Wheatstone in 1832 (first published in 1838). It is a simple instrument for presenting any two pictures separately to the two eyes. Normally these pictures are (3-D) stereo-pairs, made with two identical cameras separated by the distance between the eyes, to give the disparity which the brain normally uses to give stereo-depth vision. Wheatstone's instrument used two mirrors for viewing the stereo drawings or later photographs (Figure 3.23).

Retinal rivalry

If rather different pictures are presented to the two eyes in a stereo-scope, curious, distinctive effects occur. Parts of each eye's picture are successively combined and rejected, in what is known as 'retinal rivalry'. Rivalry also generally occurs if different colours are presented to the two eyes, though fusion into mixture colours is possible,

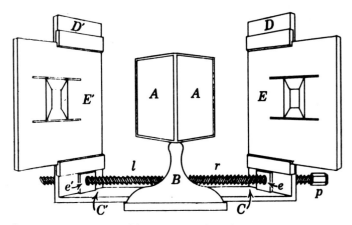

3.23 Wheatstone's stereoscope. (a) His original (pre-photography) instrument. (b) His drawings, shown opposite, appear vividly in three dimensions when viewed in a stereoscope—or as red–green anaglyphs (see Plate 1).

especially when shared contours are fused by the eyes. Subtle rejection from 'non-corresponding' retinal regions must be taking place all day long in normal perception; when we look in the middle distance, much nearer and much further objects are too separated, horizontally, for fusion by the brain to be possible. This was investigated by P. L. Panum (in 1858) who found that the limit to stereoscopic fusion—Panum's limit—is about one degree, which is continually exceeded for objects at various distances in normal viewing. This rejection of images which the brain fails to combine is a remarkable, entirely unconscious mechanism. Sometimes why, and how, we don't see can be as important and interesting as what we do see!

A great deal has been discovered over the last few years of how the brain generally combines the images of the two eyes, using the small disparity differences to compute stereoscopic depth. We now know that most cells in the visual cortex respond to stimulation from corresponding points of the two retinas. These are called binocular cells. Presumably corresponding retinal points—as related by disparity to distance—are indicated by the firing or non-firing of these binocular cells.

Julesz's random dots

Until recently it was assumed that stereoscopic vision always functions by binocular comparison between edges of objects. In what has turned out to be an unusually important technique, Bela Julesz, of

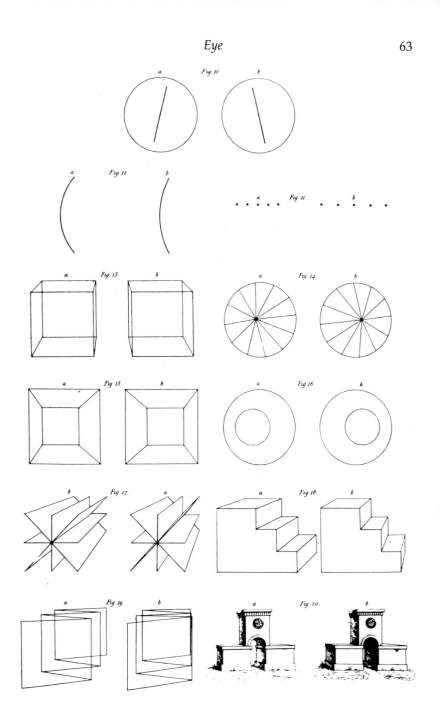

3.23 (b)

the Bell Telephone Laboratories in America, has shown that lines
or borders are not needed. Julesz generated pairs of random dot
patterns, with a computer, arranged so that for each dot in the pattern
shown to one eye, there is a corresponding dot for the other eye
(Plate 2). When groups of corresponding dots are slightly displaced
horizontally, the displaced dots are fused with their corresponding
dots of the other eye's field—and depth is seen. Regions of horizontally displaced random patterns stand out (or sink behind) the rest,
and may be given quite complex three-dimensional shapes. This
shows that cross-correlation of points, even where there are not contours, can give stereoscopic vision. Many beautiful examples will be
found in Julesz's *Foundations of cyclopean perception* (1971).

Is this all there is to stereo vision? As usual, there is more to it! First,
it is not entirely clear why some dots, but not others, are accepted as
'corresponding'. This needs some kind of global decision. Further,
illusory non-existent contours can fuse and stand in or out in stereoscopic depth (see Chapter 10).

A few people do not perceive depth from the Julesz dot figures,
though they do from ordinary line or picture stereograms. Also, for
normal observers, if there is no brightness contrast—but only colour
contrast—the dot figures do not give depth, though depth is still seen
for lines. So there may be two brain mechanisms for stereo vision.

Range-finder convergence linked to stereo disparity

There is a subtle connection between range-finder convergence and
stereo vision. The *convergence* of the eyes for near or distant objects
affects the depth signalled by the *disparity* differences between the
images. A given disparity is scaled to give greater depth for distant
objects. If this did not occur, distant objects would look thinner and
closer together in depth, than when near.

Switching the eyes

When stereo pictures are switched over, so that the right eye receives
the left eye's picture and vice versa, we should see reversed depth.
Interestingly, this does not always happen. It is also possible to switch
the eyes, or increase or decrease their separation, while viewing
normal objects, (Figure 3.24).

Depth reversal occurs in pseudoscopic vision—when the eyes are
switched over—except when reversed depth would be highly
unlikely. Thus faces will hardly reverse at all (see Figure 10.11).
Hollow faces are too improbable to be accepted. A hollow mask will
look like a normal face, until approached very closely with both eyes

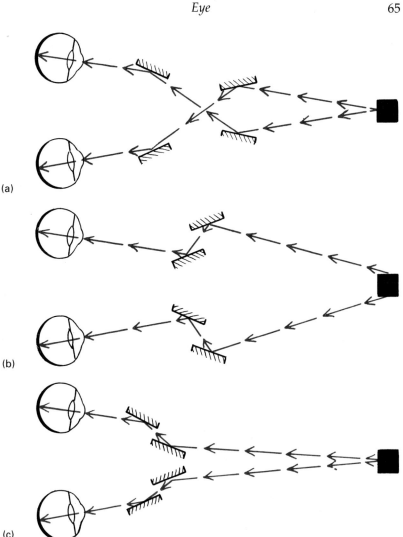

(a)

(b)

(c)

3.24 Moving the eyes with mirrors. (a) A *pseudoscope* reverses disparity. (b) A *telestereo-scope* increases the separation of the eyes, giving increased disparity. (c) An *iconoscope* reduces the effective separation of the eyes, minimizing stereo. It is possible to reduce disparity to zero. By changing the angles of the mirrors, convergence can be changed without affecting disparity. All these arrangements are useful for studying stereo and vergence depth perception, and are interesting to play with.

open. A stereo photograph of a face will refuse to appear hollow in a pseudoscope. So stereopsis is a cue to depth which may be rejected. It may be rejected by the power of knowledge (for example of faces), or by competing depth cues such as perspective (for example when viewing a stereo perspective drawing in a pseudoscope).

This shows that we must never ignore the processes of the brain on vision. The brain must 'know' which eye is which, otherwise stereoscopic depth perception would be ambiguous, and reversal of the pictures in a stereoscope (or the normal world in a pseudoscope) would have no effect. But oddly enough, when the light is cut off to one eye it is virtually impossible for us to say which eye is doing the seeing. Although the eyes are fairly well identified for the depth mechanism, this information is not available to consciousness. So experiments are needed to find the brain's contributions. Its enormous contribution here is rejecting highly unlikely objects, such as hollow faces. This shows how the 'top-down' power of knowledge can affect (and sometimes mislead) vision.

4

Brain

It has not always been obvious that brains are involved in thinking, memory, sensation, or perception. In the ancient world, including the great civilizations of Egypt and Mesopotamia, reaching back five thousand years, the brain was regarded as an unimportant organ, because in death it is bloodless and in life it is seldom felt by its owner. The mind was associated with the stomach, the liver or the gall bladder, and especially with the heart which is clearly responsive to emotion and effort. Echoes linger from these ideas in modern speech, in words such as 'phlegmatic', 'gall', 'choleric', as well as 'heartless'.

When the Egyptians embalmed their dead they did not keep the brain (which was extracted through a nostril), though the other organs were preserved in special Canopic jars placed beside the sarcophagus. As in death the brain is almost bloodless, it seemed ill-suited as the receptacle of the vital spirit. The active pulsing heart seemed to be the seat of life, warmth, and feeling, rather than the cold, silent brain, locked privily in its sunless box of bone.

The vital role of the brain for controlling the limbs, for speech and thought, and perception became clear from the effects of accidents, when the brain was damaged. Later, effects of small tumours and gunshot wounds gave specific information which has been followed up and studied in great detail. The results are of the greatest importance to brain surgeons, for while some regions are comparatively safe others must not be disturbed or the patient will suffer grievous loss. It is not, however, a simple matter to infer from loss of behaviour, or of memory or perception, just what a damaged or missing region of the brain was doing before the accident, or the operation. Removing a component from a radio may make it howl, but it does not follow that this component was a 'howl suppresser'. The circuit as a whole may be so changed that the remaining components produce this new effect, as they function differently. We need to understand this new circuit to explain what has happened. In general, removing parts is useful for isolating separate systems (as for the mechanisms for each note of a piano). But within an interacting system it is very hard to interpret results of accidental or deliberate damage or ablation.

It turns out that much of the brain—and especially for the areas involved in visual processing—is organized as separate, though inter-connected, *modules*. These can be isolated for study in various ways, including by ablation. Recording from individual brain cells has provided essential new information since almost the start of the twentieth century. Recently, PET (positron emission tomography) scans and NMR (nuclear magnetic resonance) scans have made it possible to see which regions of the brain are especially active under various conditions, such as while reading, seeing, or imagining. In general, the same regions of brain are active for seeing things and for imagining things visually. But now we have rushed ahead.

The brain has been described as 'the only lump of matter we know from the inside'. From the outside, it is a pink-grey object about the size of two clenched fists. Some parts particularly associated with vision are shown in Figure 4.1. It is made up of 'white' and 'grey' matter. The white matter contains association fibres, connecting the many kinds of cell bodies which form the grey matter of the brain. The 'high-level' activity takes place in the outer layers—the cerebral cortex—and about half the cortex is associated with processing for vision.

Through its evolution the brain has grown from the centre, which is primarily concerned with 'life support' systems such as breathing, and with emotion. The surface cortex (Latin 'bark') is curiously convoluted, in the higher animals especially humans. It is largely concerned with motor control of the limbs and with the senses. From experimental data it is possible to build maps relating regions of the cortex to groups of muscles, and to touch—producing the bizarre 'homunculus' (Figure 4.2).

Each nerve cell has a cell body and a long thin process (or axon) which conducts impulses from the cell. The axons may be very long, some extending from the brain down the spinal cord. The cell bodies also have many finer and shorter fibres (the dendrites) which accept signals. (Figure 4.3). The cells, with their interconnecting input den-drites and output axons, sometimes seem to be arranged randomly; but in some regions they form distinct patterns, indicating ordered connections. In the visual cortex they are arranged in layers, hence the term *striate cortex* for the primary visual region.

The nerve signals are in the form of electrical pulses (*action potentials*), which are the brain's only input and its only output. They

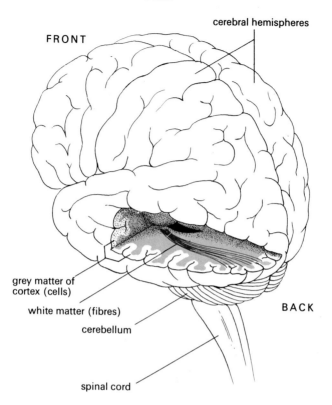

FRONT

cerebral hemispheres

grey matter of
cortex (cells)

white matter (fibres)

cerebellum

BACK

spinal cord

4.1 The primary visual region is the striate cortex (or area striata) of the brain at the back (occipital cortex, shaded grey in this diagram). Stimulation of small regions produces flashes of light in corresponding parts of the visual field. Stimulation of surrounding regions (visual association areas) produces more elaborate experiences.

depend on alteration in the ion permeability of the cell membrane (Figure 4.4). At rest, the inside of an axon is negative with respect to the surface; but when a disturbance occurs—for example when a retinal receptor is stimulated by light—the centre of the fibre becomes positive, initiating a flow of current which continues down the nerve as a wave. It travels much more slowly than electricity along a wire: in large fibres it travels at about 100 metres per second, and in the smallest fibres at less than one metre per second. The thick high-speed fibres have a special fatty coating—the myelin sheath—which insulates the fibres from their neighbours and also serves to increase the rate of conduction of the action potentials.

The low rate of travel, first measured by Hermann von Helmholtz, in 1850, when he was 29, came as a great physiological surprise.

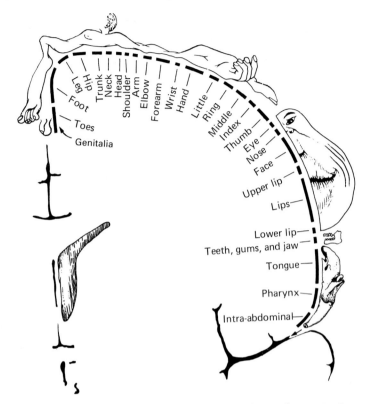

4.2 A homunculus, showing how much cortex is devoted to touch sensation from various regions of the body. Note the huge thumb. Different animals have different homunculi, corresponding to the sensory importance of various parts of the body.

Helmholtz's professor, Johannes Muller (1801–58), the founder of modern physiology, thought it must be faster than the speed of light and forever unmeasurable. How was it measured, before sophisticated electronics? (See box, page 71).

The travel time in peripheral nerves only accounts for part of the human reaction-time of about a twentieth of a second, for (as Helmholtz realized and measured) there is also delay in the switching time of the synapses of the brain, processing and routing sensory and motor signals. These time measures have been used to tease out different brain processes.

The *area striata* is sometimes known as the 'visual projection area'. When a small part is stimulated with an electrode, a human patient reports a flash of light. Upon a slight change of position of the stimulating electrode, a flash is seen in another part of the visual field.

Measuring the speed of nerve signals with a stop watch

This can be done in a simple experiment. Arrange a row of perhaps 20 people on closely spaced chairs (they can be in a circle) with their hands poised to touch their neighbour as quickly as possible. Get each to touch their neighbour's upper arm as soon as they feel a touch from their neighbour on the other side. Time the cumulative delay with a stop watch. Repeat say 10 times. Try the same again—but touching wrists.

The difference in path-length from shoulder to wrist is about 50 cm. So with 20 people there is an extra 10 metres of nerve. This is enough, averaging several trials, to get a reasonable measure of the speed of nerve impulses. The time each person takes to move their finger does not matter provided it is the same in both conditions.

Stimulation of regions surrounding the striate area give more elaborate sensations than flashes of light: visual memories, even complete scenes, coming vividly before the eyes.

There is rough mapping of the retina at the first stages of processing, but this is soon lost as processing continues, in various specialized modules. There is separate parallel processing of form, movement, colour, and probably many other visual 'dimensions'. This is a recent counter-intuitive discovery, that there are many separate processing modules, which somehow pool their results to give a single, generally consistent perception. How they come together is not yet known.

There are also separate 'channels' for the fine detail and broad brush strokes of a scene. Visual scientists speak of 'spatial frequencies' by analogy with temporal frequencies of sound. When we look at a painting, the texture of the picture (literally the brush strokes) is processed by a different spatial frequency band from the depicted objects; but when the texture and objects are similar in scale they become (sometimes deliberately) confused. Processing of form and colour is largely separate physiologically, colour being by 'broad brush' low spatial frequency channels, operating quite slowly.

It is an interesting question how colour generally remains 'registered' to luminance contours, or edges, though sometimes colour does 'bleed' through gaps. Pictures with colour contrast but no brightness contrast (isoluminance, or equiluminance) look very peculiar. There is a curious instability, and with slight movement the colours separate. A face made up of equal brightness dots, of say red and green, does not look like a face. It is a meaningless shape.

It has been discovered by the psychologist Anne Treisman that with

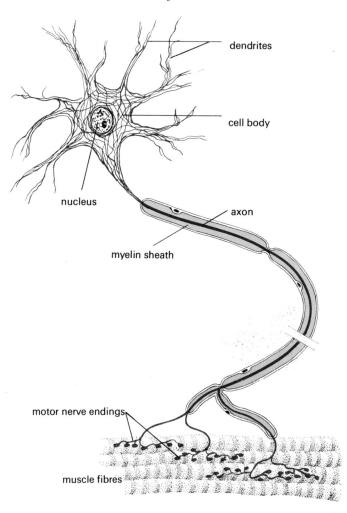

4.3 A nerve cell. The cell body has a long axon, insulated by its myelin sheath. The cell body accepts information from the many fine dendrites, some of which tend to make a cell fire, while others inhibit signals.

normal contrasts, colour and form can become dissociated. When letters of various colours are flashed up, the subjects sometimes wrongly attach a colour to a letter. She believes that the linkage occurs later, perhaps by a top-down process involving meaning and attention, because objects of known colour are seldom if ever transposed. These are important subjective effects.

Sophisticated techniques for studying the nervous system physio-

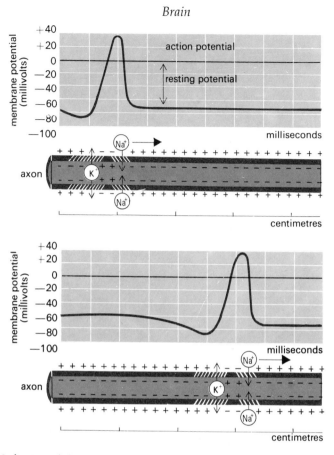

4.4 Mechanism of electrical conduction in nerve. Alan Hodgkin, Andrew Huxley, and Bernard Katz discovered that during an action potential, sodium ions pass to the inside of the fibre, converting its standing negative charge to positive. Potassium ions leak out, restoring the resting potential. This can happen up to several hundred times a second, transmitting spikes of electricity which run along nerve as signals—by which we know the world through perception (afferent nerves), and command behaviour (efferent nerves) by activating muscles.

logically allow electrical activity of individual cells to be recorded. Different regions may be stimulated electrically to evoke responses and, in patients undergoing brain operations, sensations. Among the most remarkable of recent discoveries are specialized brain modules for recognizing hands (discovered by Charles Gross), and others (especially investigated by David Perrett) for faces. They become active only when hands, or faces, are seen. This special processing of vitally important objects fits the clinical finding that with brain

4.5 The Thompson Effect. The eyes and mouth have been cut out and turned upside down. Try turning the book upside down: one hardly notices the rotation of eyes and mouth. (Modified from Thompson 1980.)

damage there can be selective loss just of face recognition, though fortunately this is rare.

What might be called a topsy-turvey-face (The Thompson effect) is suggestive. If the mouth and eyes are cut out of a picture of a face, turned upside down and replaced, the face looks bizarre, even frightening. If now the whole face is turned upside down it looks almost normal. That is, the rotation of the mouth and eyes is hardly noticed (Figure 4.5). This suggests that it is these features especially that are processed by the face module.

Loss of a region of brain may result in behaviour changes related to the region of damage—though, as we have said, there can be problems with this type of analysis. Another way of approaching this question is to investigate effects of drugs or chemicals applied directly to the brain surface. This is becoming an important technique to establish whether new drugs have unpleasant psychological side-effects, as well as for changing the state of the brain to discover functions of various regions.

These techniques, together with anatomical studies of how different regions are joined by bundles of fibres, have made it clear that different parts of the brain are engaged in very different functions, and these are becoming clearer. But ultimately we need to know more of the principles by which the brain works to appreciate their significance. This requires experiments to be interpreted by conceptual models of what could be happening. It may be necessary to invent imaginary brains—by constructing functional machines and writing computer programs to perform perhaps much like biological systems. In short, we may have to simulate to explain; though simulations are never complete or perfect. So far we have no machines which approach the brain in thinking or seeing. Nevertheless, important principles may be suggested. If the brain were strictly unique, then we could hardly hope to find general explanations, for science thrives on analogies and is blinded by the unique. This is a controversial matter, especially for the issue of consciousness. We may watch out for developments in machine perception and intelligence to illuminate the brain and mind from the outside—but what of inside experience?

The neural system responsible for vision starts with the incredibly complicated structures of the retinas (see Chapter 3). These are outgrowths of the brain, and contain typical brain cells and interconnections, as well as specialized light-sensitive detectors. Each retina is effectively divided down the middle: the optic nerve fibres from the inner (nasal) halves cross at the chiasma, while fibres from the outer halves do not cross (Figure 4.6). In addition to the area striata, fibres project to a second area—the superior colliculus—which gives cruder mapping, and provides command signals to move the eyes. This is evolutionarily more ancient: it seems that the sophisticated feature analysis of the area striata has been grafted on to the primitive but still useful—though not conscious—vision of the colliculus, as the cortex developed through evolution.

The brain has two hemispheres, which are each more-or-less complete brains, joined by a massive bundle of fibres, the corpus callosum, and the smaller optic chiasma. On their way from the chiasma, the optic tracts pass through a relay station in each hemisphere, the lateral geniculate bodies (or nuclei). Interestingly, they receive more fibres 'downwards' from higher centres than from the eyes. This is an anatomical basis for higher centres modulating or adding to visual signals, giving meaning to retinal images.

Among the most important discoveries of recent years is the finding of two American physiologists, David Hubel and Torstin Wiesel, who recorded activity from single cells of the area striata of the cat's brain while presenting its eyes with simple visual shapes. These were generally bars of light, projected by a slide projector on a screen in front of the cat. Hubel and Wiesel found that some cortical cells of the first stage of processing (V1) were only active when a bar was presented at

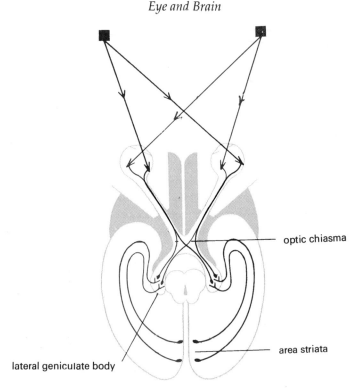

optic chiasma

area striata

lateral geniculate body

4.6 The eyes' pathways and processors. The optic nerve divides at the chiasma, the right half of each retina being represented on the right side of the occipital cortex, the left side on the left half. The lateral geniculate bodies (or nuclei) are relay stations between the eyes and the visual cortex.

a certain orientation (Figure 4.7). At other angles they were 'silent'. Different cells would respond to different orientations. Cells deeper in the brain responded to more generalized characteristics, and responded to these characteristics no matter which part of the retina was stimulated. Other cells responded only to movement, and some to movement in only a single direction (Figure 4.8). These findings are of the greatest importance, for they show that specific mechanisms in the brain select certain features of objects. Perceptions are built up from combinations of these selected features.

The striate cortex is organized not only in the clearly seen layers parallel to its surface, but also in functional blocks running through the layers. An electrode pushed through the layers, at a right angle to the surface, picks up cells which all respond to the same orientation. If the electrode is inserted about half a millimetre away, the critical

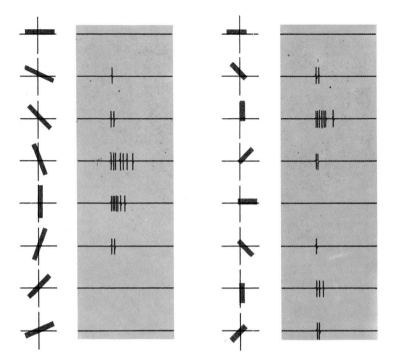

4.7 Responses to orientation. Hubel and Wiesel's records from single cells in the visual cortex (area V1) of the cat. A narrow bar was presented at various orientations. A single cell in the brain fires strongly at a certain orientation, more weakly when this orientation is changed a little, and not at all for quite different angles. This is shown by the spikes of the electrical records.

orientation is different. Each block (as Colin Blakemore puts it) of 'like-minded' orientation detectors is called a column. Various sizes of retinal image features, velocities, and (in monkeys and probably humans) colours, are represented by cells of common orientation down each functional column. Most of the cells are 'binocular'— responding to stimulation of corresponding points of both retinas. Deeper cells respond with more and more general properties, from 'simple', to 'complex', and 'hypercomplex' cells.

Channels for form and colour

It is becoming ever more clear that the brain processes visual characteristics in specialized 'modules'; with different neural channels for form, movement, stereo depth, colour, and so on. Colour came late in

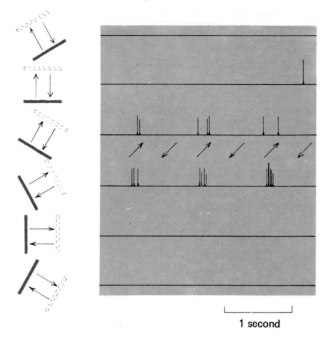

1 second

4.8 Responses to moving objects. Selected single brain cells fire when a line, or bar, is moved across the cell's receptive field of the retina. The electrical record shows that this particular cell fires with only one direction of movement.

mammalian evolution—practically non-existent before primates—and its brain processes are late add-ons and largely separate from form perception. Thus what are called 'blobs' from their appearance (in upper layers of area V1) have cells specialized for colour, with almost no response to form or movement. Cells for colour and form are in thin alternating layers. It has been found by David Hubel and Margaret Livingstone that in the relay station between the eyes and the cortex, the lateral geniculate nucleus (LGN), there are large (magno) and small (parvo) cells, serving colour and form respectively. The magno system is colour blind. It signals form and movement (to Area V5) and works much faster than the colour-serving parvo system. So now we know why flicker by alternating between colours of equal brightness, is only seen up to about 12 hertz; though brightness flicker is seen up to 50 hertz or higher. As well as illuminating how the brain works, facts such as these are of practical importance for cinema and television engineering.

Gradually, the organization of the 'visual cortex' is becoming clear, though just how this is related to seeing the complex shapes of objects remains unclear. Perhaps we won't fully understand the visual brain before we design and make a machine with sophisticated vision.

Seeing objects by machine and brain

There have been several attempts, over the last thirty years, to program computers to recognize objects from images of a television camera. Initially a small range of objects was used to simplify the problem, such as cuboid blocks, pyramids, and cylinders. The pioneering achievement for Terry Winograd, in 1972, was to construct such an imaginary world in a computer, which could describe and even invent new words for what it saw within itself.

Perhaps more closely related to eyes and brains was David Marr's well known scheme suggesting a series of stages of processing from the eyes 'upwards', though with little or no call on knowledge deployed 'downwards'. The scheme starts with measures of local brightnesses, giving the 'primal sketch' in computer terms, leading to the charmingly named '$2\frac{1}{2}$-D sketch', which is a preliminary organization of flat images into tentative three-dimensional descriptions of arrangements of objects in space. It turns out that edges are very important; but it is often hard to decide when edges belong to a single object, or are shared. This problem is made tractable by setting restraints for what is likely, from common shapes of objects. Also, Marr accepted that most objects can be constructed from a few simple shapes, such as cylinders; so his system describes complex objects in terms of a few unit shapes. Actually this idea goes back to earlier artists: Cézanne wrote in a letter in 1904, 'Treat nature by means of the cylinder, the sphere, the cone, everything brought into proper perspective.' This is useful for art students; they too learn to segment three dimensional objects, especially the human form, into what the AI researchers D. D. Hoffman and W. A. Richards call 'concave discontinuities'. Such tie-ups between machine object recognition on the one hand, and human vision and art on the other, may be deeply suggestive for linking how we see with how we may build seeing robots; but whether robots will ever appreciate art is an open question!

David Marr chose object-unit shapes convenient for computer analysis, but are these appropriate for human vision? More recently, Irving Biederman has designed experiments looking for basic shapes that might be used for human vision, which he calls 'geons'. His

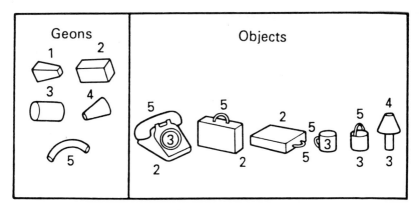

4.9 Biederman's geons—unit object shapes—fitted to some common objects.

ingenious psychological experiments distinguished between visual processing and conceptual understanding of objects by, for example, asking whether a different view of the same piano 'primes' recognition from a different view as well as, or better than, a similar view of a different kind of piano—upright to grand. Biederman provides evidence for geons and for the importance of conceptual understanding for human vision.

Marr's unit descriptions are based on a particular viewpoint. Biederman's geons are defined by properties remaining essentially the same (invariant) over a wide range of views. So for Biederman, though not for Marr, object recognition can be achieved directly from the two-dimensional primal sketch, with no need to construct an internal three-dimensional shape though knowledge may be needed. Figure 4.9 shows Biederman's geons fitting common objects.

These schemes are carried out with digital computers; but it is far from certain that the brain works by digital computing. It now seems more likely that the brain has many modules of self-adapting nets (page 82). If so, Marr's (and others') suggested computer algorithms for vision may be descriptions of what operations the nets carry out, rather than what is actually happening in brain. This distinction is important for relating physiological activity to processes of vision. It is easy to confuse descriptions with what (perhaps) is being described.

The actual physiology is quite unlike the many high-speed switches needed for digital computing, and the brain seems to be self-adapting rather than programmed with computer algorithms. Current work on artificial neural nets (though this started forty years ago) looks promising for a more realistic account of how the brain works.

Imaging brain functions

Recently it has become possible to see which regions of brain are active, in animals and humans, without doing damage or causing undue distress. PET and NMR scans have revolutionized brain research as well as being invaluable tools for the clinician. The patient or experimental subject puts his or her head inside what looks like a large tumble-drier, and unfortunately the machine makes a similar noise as the mechanism rotates in the same way. In PET scanning the subject is injected with a very weakly radioactive compound and the scanner detects alpha particles emitted from the brain's blood supply. It has been known for a hundred years that blood supply increases locally with raised activity, and PET scanning now makes it possible to record these increases. A computer builds up a map of the activity, as three-dimensional thin slices through the brain.

In a recent experiment using scanning to detect brain activity, subjects looked at black and white drawings of an elephant. They were asked to imagine either the elephant's colour, or its behaviour. When they imagined colour (grey), regions close to the known colour processing region (area V4) became active. When they imagined the elephant's behaviour, regions close to the movement processing (area V5) became active (Figure 4.10). What is interesting, is that these brain changes occurred with no change in the visual stimulus. All that changed was what the subjects imagined. It turns out that essentially the same regions are active for vision of colour or movement or form,

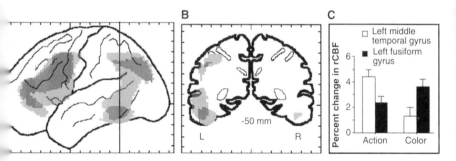

4.10 Brain activity associated with imagining colour and movement. When subjects imagined the colour of an elephant, brain regions associated with colour vision became active (green). When they imagined movement, movement regions lit up (grey). As the input was unchanged, this reveals physiological bases of cognition. (From Martin, A., *et al.* (1995) Discrete cortical regions associated with knowledge of colour and knowledge of action, *Science*, **270**, 102, © 1995 American Association for the Advancement of Science.)

as for imaging colour or movement or form. In dreaming, the same regions are active as for waking vision, or hearing, or movements of the limbs. The same is so for the inner voices and other hallucinations of schizophrenia. So, at last, cognitive brain processes are being identified with experiences and actions. There are limits to the available resolution in time and place, but these non-invasive techniques are providing new knowledge, and promise a great deal more in the near future for understanding how cognition is related to physiology. This identifies where to look for what is going on. We still have to discover just what is happening in the brain when we see with the eyes, or imagine with the mind's eye, with consciousness.

The brain is often compared to a computer—but what kind of computer? Computers come in two very different kinds: analogue and digital. A graph is analogue, allowing, for example, average height of children to be read off according to age. Digital computing involves going through the steps (algorithms) of calculation or logic. Analogue systems avoid actually having to go through the steps, so they are not truly computers—they avoid computing. Compared with digital computers they are fast, but somewhat inaccurate. Digital computing requires many switches which must work fast and with high reliability. The brain does not look like this. It looks much more like a collection of analogue *neural nets*, somewhat as the Canadian psychologist Donald Hebb conceived in his book *Organization of behaviour*, in 1949. Artificial intelligence (AI) is returning to this kind of view of the brain. The most influential account of vision as being given by digital algorithms, the brain being regarded as a digital computer, is David Marr's book *Vision* (1982), but this approach is now being questioned.

The key notion of nets is that cells become more active (or more conductive) as they are stimulated more often. Attempts were made early on to construct seeing machines with artificial nets. Most famous was Frank Rosenblatt's Perceptron. With excitatory and inhibitory connections (synapses), it could begin to generalize patterns. There were two layers of artificial neurones, with every input (retinal receptor) connected randomly to every output. With the cells adapting to frequency of use, this system gradually became structured: its internal patterns representing external regulations of its world. It could give some generalizations of patterns. However, in 1969 Marvin Minsky and Seymour Pappert showed that there must be severe limitations in Rosenblatt's two-layer Perceptrons. Largely as a result of their criticisms, the idea was dropped. Interest also declined because of the remarkable power and flexibility of the new digital computers.

If the interest in Perceptrons had not died, it might have been real-

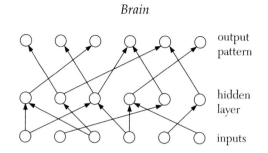

4.11 A three-layer neural net. The lower nodes are inputs, the upper nodes are outputs. The central hidden layer develops mind-like generalizations.

ized much sooner that adding a layer of 'hidden units' between its inputs and outputs can make a fundamental difference (Figure 4.11). These allow inner patterns to develop, which are not driven by the input or closely related to the output. Protected from input and output, they are hidden and secret—rather as mind is hidden and secret.

These 'hidden units' can abstract and learn, discover and create generalizations. They can recognize patterns, even from only a small part being visible. Perhaps suggestively, they need periods of rest to sort themselves out—to dream! Like the brain, and unlike digital computers, they go on working even when a large proportion of their components are destroyed.

Terence Sejnowski, in California, has devised a net that can learn to read English, very much as children do, starting with random babbling. His artificial net is faster than babies at learning from examples, though one does not known how sophisticated it can become. This is a general query about neural nets. Many work well for small problems but become inadequate when having to deal with a large number of alternatives. How best to teach artificial neural nets presents unsolved questions. Here may lie some helpful hints and perhaps significant research for human education. The essential point of self-adapting nets is that analogue interactive systems can learn from successive presentations of, for example, faces, letters, or sounds of words, to recognize new objects of the same class and build up new classes or categories of objects and situations by teaching themselves. Is this where the answer lies for how the brain works?

5

Seeing brightness

There is supposed to be a tribe of cattle breeders who have no word for green in their language but have six words for different shades of red. Specialists in all fields adopt special meanings for their own use, and generally name more distinctions than are usually seen or at least noticed. Before embarking on a discussion of brightness and colour we should stop for a moment to sharpen some words—as a carpenter might stop to sharpen his chisel before attempting delicate work.

We speak of physical *intensity* of light entering the eye giving rise to apparent or seen *brightness*. Physical intensity of light may be measured by various kinds of photometer, including the familiar photographer's exposure meter; but brightness is an experience, which is very hard and perhaps impossible to measure. We believe we know what another person means when he says: 'What a bright day!' He means, not only that he could take photographs with a slow film in his camera, but also that he experiences a dazzling sensation. This sensation is roughly, but only roughly, related to the intensity (also called 'luminance') of the light entering the eye.

When talking technically about colour vision we do not generally talk of 'colours' but rather of 'hues'. This is simply to avoid the difficulty that 'colours' are apt to mean sensations to which we can give a specific name such as 'red' or 'blue'. We thus speak technically of 'spectral hues' rather than 'spectral colours', but this is not always necessary. The distinction between physical intensity and seen brightness is more important.

Another important distinction to be made is between *colour as a sensation* and *colour as a wavelength* (or set of wavelengths) of light. Strictly speaking light itself is not coloured: it gives rise to sensations of brightness and colour, but only in conjunction with a suitable eye and nervous system. Although Sir Isaac Newton was quite clear on this in his *Optics* of 1704, the technical language is somewhat confused on this matter. We do speak sometimes of 'coloured light', such as 'yellow light', but it should be taken to mean light which generally gives rise to a sensation described by most people as 'coloured' or specifically 'yellow'.

Without attempting to explain how physical intensities and wavelengths of radiation give rise to different sensations (and ultimately we do not know the answer), we should realize quite clearly that without life there would be no brightness and no colour. Before life came, especially higher forms of life, all was invisible and silent though the sun shone and the mountains toppled.

The simplest of the visual sensations is brightness. It is impossible to describe the sensation. A blind man knows nothing of it; yet to the rest of us, reality is made up of brightness and of colour. The opposed sensation of blackness is as powerful—we speak of a 'solid wall of blackness pressing in on us'—but to the blind this also means nothing. The sensation given to us by absence of light is blackness; but to the blind it, it is nothingness. We come nearest to picturing the world of the blind, who have no brightness and no black, by thinking of the region behind our heads. We do not experience blackness behind us: we experience nothing, and this is very different from blackness.

Brightness is not just a simple matter of the intensity of light striking the retinas. The brightness seen by a given intensity depends upon the state of adaptation of the eyes, and also upon various complicated conditions determining the contrast of objects or of patches of light. In other words, brightness is a function not only of the intensity of light falling on a given region of the retina at a certain time, but also of the intensity of the light that the retina has been subject to in the recent past, and of the intensities of light falling on other regions of the retina.

Dark–light adaptation

If the eyes are kept in a low light level for some time they grow more sensitive, and a given light will look brighter. This 'dark adaptation' is rapid for the first few seconds, then slows down. The rod and cone receptor cells adapt at different rates: cone adaptation is completed in about seven minutes, while rod adaptation continues for an hour or even more. This is shown in Figure 5.1 which shows the two adaptation curves—one for the rods, the other for the cones. It is as if we have not one but two retinas, lying intermingled in the eye.

The mechanisms of dark adaptation are beginning to be understood in detail, largely through the ingenious and technically brilliant

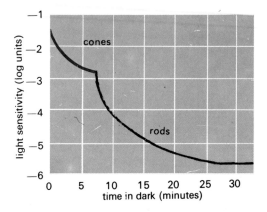

5.1 Increase in sensitivity of the eye in the dark, known as dark adaptation. The *red* curve shows how the cone cells adapt, while the *black* curve shows rod adaptation, which is slower and proceeds to greater sensitivity. In dim light only the rods are functional, while they are probably inhibited in brighter light by the active cones.

experiments of the British physiologist W. A. H. Rushton at Cambridge. It was suggested many years ago that adaptation is due to regeneration of the visual pigments of the eye bleached by light—this bleaching in some unknown way stimulating the receptors to transmit electrical signals to the optic nerve. The photochemical rhodopsin was extracted from the frog's eye, and its density to light measured during bleaching and regeneration, and compared with human dark adaptation curves. The two curves are shown together in Figure 5.2, and indeed they do correspond very closely, suggesting a strong connection between the photochemistry of rhodopsin and the changing sensitivity of the rod eye. It would also seem that brightness must be related to the amount of photochemical present to be bleached. What Rushton has done is to measure the density of the photochemical in the living eye, during adaptation to darkness. The technique is, essentially, to shine a brief flash of light into the eye and to measure the amount of light reflected from it with a very sensitive photocell. At first it seemed impossible to do this for the human eye because so little light remains to be reflected after the almost complete absorption by the photochemicals and the black pigment lying behind the receptors. So a cat's eye was used, the reflecting layer at the back, the tapetum, serving as a mirror to reflect light to the photocell. The method worked with the cat's eye and Rushton then succeeded in making it sufficiently sensitive to detect and measure the very feeble light reflected from the human eye. He found that there is bleaching of the photochemicals with adaptation, though this was much less than

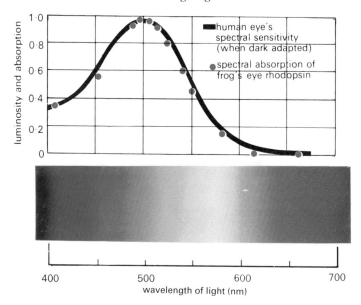

5.2 The chemical basis of vision. The curve in *black* shows the sensitivity of the dark adapted human eye to various wavelengths of light. The *red* dots show the amount of light over the same range of wavelengths absorbed by the photochemical rhodopsin in the frog's eye. The curves are substantially the same, indicating that the human eye (when dark adapted) functions by absorption of light by the same photochemical.

expected. He then detected the three colour-sensitive pigments in this way, obtaining results broadly confirmed by microscopic spectral absorption measurements of individual cone cells.

Contrast

Another factor which affects brightness is the intensity of surrounding areas. A given region generally looks brighter if its surroundings are dark, and a given colour looks more intense if it is surrounded by its complementary colour. This is no doubt related to the cross-connections between the retinal receptors. Contrast enhancement seems to be tied up with the general importance of borders in perception. It seems that it is primarily the existence of borders which is signalled to the brain. Little information is needed on regions of constant intensity, and the visual system extrapolates information between borders, which no doubt saves a lot of information-handling. Although the phenomena of contrast and enhancement of borders are no doubt mainly because of retinal mechanisms that are due to cross connections between cones, which effectively sharpen the signalling of borders (by lateral inhibition), there also seem to be more central contributions. This is brought out in Figure 5.3. There is more brightness difference in the grey ring when a fine thread is placed across it, marking the division of the background. The contrast is greater when the figure is interpreted as two separate halves than when it is regarded as all one figure, which perhaps suggests that 'central' processing is occurring.

Something of the subtlety of the human brightness system is shown by Fechner's paradox. This is as follows. Present the eye with a small

5.3 Simultaneous contrast. The part of the grey ring against the black appears somewhat lighter than the rest of the ring against the white background. This effect is enhanced when a thread is placed across the ring, along the black–white junction.

fairly bright source: it will look a certain brightness, and the pupil will close to a certain size when the light is switched on. Now add a second, dimmer light. This is placed some way from the first, so that a different region of retina is stimulated. What happens? Although the total intensity has increased, with the addition of the second light, the pupil does not close further as one might expect. Rather, it opens—to correspond to an intensity between the first and the second light. It is evidently set not by the *total*, but by the *average* illumination.

Try shutting one eye, and note any change in brightness. There is practically no difference, whether one or two eyes receive the light. This is not so, however, when small dim lights are viewed in surrounding darkness: they *do* look considerably brighter, with two eyes than with one. This phenomenon is not understood.

Brightness is a function of colour. If we shine lights of different colours, but the same intensity, into the eyes, the colours at the middle of the spectrum will look brighter than those at the ends. This is shown in Figure 5.4, the curve being known as the spectral luminosity curve. This is of some practical importance, for if a distress signal light is to be clearly visible it should be of a colour for which the eye is maximally sensitive—in the middle of the spectrum. It is complicated by the fact that, although the sensitivity curves for rods and cones are similar in general shape, the cones are most sensitive to yellow, while the rods are most sensitive to green. The change with increasing intensity is known as the Purkinje shift, after the Austrian physiologist Johannes Evangelista Purkinje (1787–1869), who was an extremely good observer of 'entoptic' (within the eye) phenomena.

The luminosity curve tells us nothing much about colour vision. It is sensitivity to light plotted against wavelength of light, but with no reference to the colours seen at each wavelength. Animals without colour vision show a similar luminosity curve.

It seems that, although there are photochemical changes associated with adaptation to light, there are also several additional mechanisms at work, these being not photochemical but neural. In particular, as the eye becomes dark adapted, it trades its acuity in space and time for increase in sensitivity. With decrease of intensity, and the compensating dark adaptation, ability to make out fine detail is lost. This is no simple matter, but it is in part due to the retina integrating over a greater area and so a greater number of receptors. There is also an increase in the time over which photic energy is integrated as the eye adapts to dim light, much as photographers use longer exposures to compensate for low light levels.

The trading of temporal discrimination for sensitivity with dark

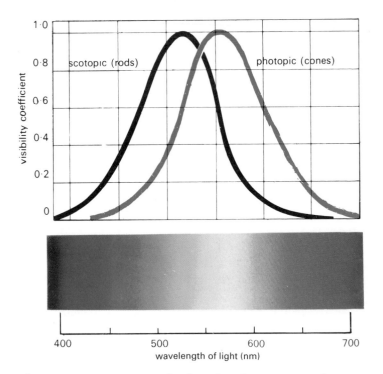

5.4 The spectral luminosity curve. This shows how the sensitivity of the eye to various wavelengths is different when the eye is light adapted. The *black* curve shows that the maximum sensitivity shifts along the spectrum when the eye is dark adapted. The *red* curve shows sensitivity when the eye is light adapted. This is the Purkinje shift, and is presumably due to changing relative contributions from the rod and cone receptors.

adaptation is elegantly (if somewhat indirectly) observed in a curious and dramatic phenomenon known as the Pulfrich pendulum effect (Figure 5.5). One remarkable feature of this is the way it was discovered, because it cannot be seen without two functional eyes and yet its discoverer was blind in one eye! The experiment is well worth trying. Take a length of string and a weight for a bob, to make a pendulum about one metre long. Swing the pendulum in a straight arc at right angles to the line of sight. View the oscillating bob with both eyes, but cover one with a dark, though not opaque, glass such as one filter of a pair of sunglasses.

You will see that the bob does not appear to swing in a straight arc, but to describe an ellipse. The ellipse may be extremely eccentric—indeed the longer axis can lie along the line of sight, though the bob is actually swinging straight across the line of sight. What causes this

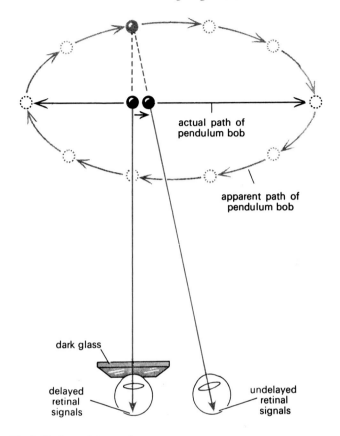

5.5 The Pulfrich pendulum. When a pendulum swinging in a straight arc across the line of sight is viewed with a dark glass (such as one filter of a pair of sunglasses) over one eye, both eyes being open, it appears to swing in an ellipse. This is due to signals from the eye which is dark-adapted by the dark glass being delayed. The bob's increasing velocity towards the centre of its swing gives an increasingly signalled disparity, accepted as stereo signals corresponding to an elliptical path.

strange effect? By reducing the light, the dark glass delays signals from this eye. These receptors take longer to respond, and the dark adaptation produces a delay in the message reaching the brain from this eye. The delay causes the affected eye to see the bob slightly in the past. As the bob speeds up in the middle of its swing the delay becomes more important, for the eye with the filter sees it in a position further and further behind the position signalled to the brain by the unaffected eye. This gives an effective horizontal shift of the moving image—as signalled—generating stereo depth. To the brain, it is as though the bob is swinging elliptically. It seems that increased delay with dark adaptation is associated with increase in temporal integrating time: as when a photographer uses a longer exposure in dim light. We see this also, and more directly, from the 'comet's tail' following a moving firework at night, as dark adaptation increases the effective exposure-time of the eye to increase its sensitivity.

Both the increase in the delay of messages from the retina to the brain, and the increase in the integrating time which this allows, have some practical significance. The retinal delay produces a lengthening of reaction time in drivers in dim light, and the increased integrating time makes precise localization of moving objects more difficult. Games cannot be played so well: the umpire calls 'Cease play for poor light' long before the spectators think it right to bow before the setting sun.

The eye's sensitivity to light

As intensity of light is increased, the rate of firing of the receptors increases, intensity being signalled by the rate of firing. Unfortunately it is not possible to record the electrical activity in the receptors of a vertebrate eye because the retina is 'inside out' so an electrode cannot be attached to receptors without doing extensive damage. By the time signals reach the optic nerve, they have been complicated by the cross-connections of the layers of nerve cells in the retina. There is however one type of eye—that of a living fossil, the horseshoe crab, *Limulus*, found on the eastern seaboard of the United States—in which the receptors are connected directly to separate nerve fibres. Figure 5.6 shows the electrical activity in a nerve fibre of a *Limulus* eye. The rate of firing of its receptors is logarithmically related to intensity.

Figure 5.6a shows a rate of firing after the eye has been in the dark for one minute. Figure 5.6b shows the firing rate increasing as the eye has been in the dark a longer time. This corresponds to our own experience of increasing brightness after being in the dark.

What happens when we look at a very faint light in an otherwise dark room? One might imagine that in the absence of light there is no activity reaching the brain, and when there is any light at all the retina signals its presence and we see the light. But it is not quite so simple. In the total absence of light, the retina and optic nerve are not entirely free of activity. There is some residual neural activity reaching the brain even when there is no stimulation of the eye by light. This is known from direct recording from the optic nerve in the fully dark-adapted

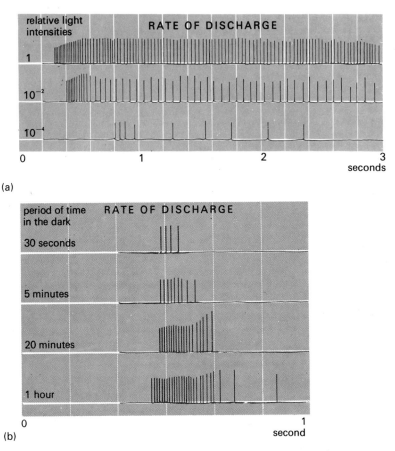

5.6 (a) Electrical activity, recorded on an oscilloscope, of a single fibre of the optic nerve of *Limulus* at three intensities of light. The rate of firing increases roughly logarithmically to the intensity. (b) The rate of firing after various durations of darkness. With increasing dark adaptation the firing rate increases, leading to an increase in apparent brightness, though the actual intensity of the light remains the same.

cat's eye, and we have strong reasons for believing that the same is true for the human and all other eyes. This continuous background of random activity is of great importance: it sets a continuous problem.

Imagine some neural pulses arriving at the brain: are they due to light entering the eye, or are they merely spontaneous noise in the system? The brain's problem is to 'decide' whether neural activity is representing outside events, or whether it is mere noise, which should be ignored. This is a situation familiar to a communications engineer, because all sensitive detectors are subject to random noise, which degenerates signals, and limits the sensitivity of detectors. There are ways of reducing harmful effects of noise, which are applied with good effect by radio astronomers (noise masks the radio sources in space just as it masks and confuses weak visual signals). The eye adopts certain measures to reduce the effects of noise, notably by increasing the duration over which signals are integrated (which we saw reflected in the Pulfrich pendulum effect) and by demanding several confirming signals from neighbouring receptors serving as independent witnesses.

One of the oldest laws in experimental psychology is Weber's Law (Figure 5.7). This states that the smallest difference in intensity which can be detected is directly proportional to the background intensity. For example, if we light a candle in a brightly lit room, its effect is scarcely noticeable; but if the room is dim to start with (lit by just a few other candles) then the added candle makes a marked difference.

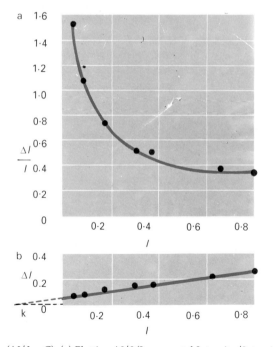

5.7 Weber's law ($\Delta I / I$ = C). (a) Plotting $\Delta I / I$ (Incremental Intensity/Intensity) against I gives a horizontal straight line over a wide range of I. But the law breaks down at low intensities (as shown here), when ΔI must be raised to be detected because of the residual neural activity, or 'noise', which is present in all communication channels. (b) The breakdown is roughly represented by adding a small constant, k, in the denominator of Weber's law ($\Delta I / I + k$ = C). See text for more details.

In fact we can detect a change in intensity of about one per cent of the background illumination. This is written $\Delta I/I = C$, where C is a constant, and ΔI is the small incremental intensity over the background intensity, I. This law holds fairly well over a wide range of background intensity but it breaks down for low intensities. If Weber's Law did hold for all values down to zero intensity, Figure 5.7a would show a horizontal straight line, indicating invariance of the just detectable differential intensity ($\Delta I/I$) over all values of I. In fact, there is a marked rise in $\Delta I/I$ as the background intensity becomes small. This breakdown is largely explained if we take into account the residual firing of the retinal cells in the absence of light. To the brain, this residual activity is exactly equivalent to a more-or-less constant dim light added to the background. We may estimate its value by plotting ΔI against I, extrapolating back past the origin and reading off the y-axis of the graph (Figure 5.7b). This gives a value for residual activity (or noise) in terms of an equivalent light intensity, which we may call k. There is evidence that this internal noise of the visual system increases with age, and is partly responsible for the gradual loss of all visual discriminations with ageing. Increased neural noise may also affect motor control and memory.

The idea that discrimination is limited by noise in the nervous system has far-reaching consequences. It suggests that the old idea of a *threshold* intensity, above which stimuli need to be if they are to have any effect on the nervous system, is wrong. We now think of any stimulus as having an effect on the nervous system, but only being accepted as a signal of an event when the neural activity is unlikely to be merely a chance increase in the noise level (Figure 5.8). This shows a patch of light serving as a background (I) on which is an added light (ΔI). These two intensities of light give rise to statistically distributed neural impulse rates. The problem for the brain is to 'decide' when a given increase is merely noise, and when it is due to the increased intensity of the signal. If the brain accepted *any* increase from the average activity, then we would often 'see' flashes of light that are not in fact present. We thus reach the idea that a statistically significant difference is demanded before neural activity is accepted as representing a signal. The smallest difference (ΔI) we can see is determined not simply by the sensitivity of the receptors of the retina, but also by the difference in neural pulse rates demanded for acceptance as a genuine signal.

When we do see flashes which are not there, these are due to the noise exceeding the demanded significance level.

The brain's 'choice' of the level above which activity is accepted is a

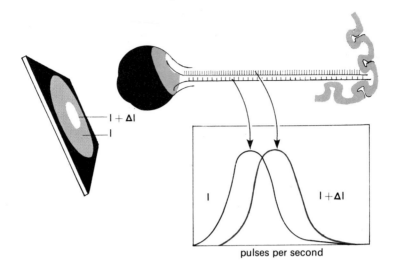

5.8 The statistical problem presented to the brain by the random firing of nerves. When the signal field ($I + \Delta I$) is being discriminated from its dimmer background (I), the pulse rates are not always different, but are distributed as in the graph. So we may 'see' a light due to noise, or miss seeing a genuine light when the rate happens to be lower than average. The brain demands a significant difference before accepting neural activity as a signal.

matter of trading reliability for sensitivity. There is evidence that this level is to some extent variable, and depends on our 'set' to mobilize physiological and psychological resources for a given situation, task, or skill. When we are particularly careful, more information is demanded and sensitivity suffers.

Intensity discrimination applies in this way throughout the nervous system. It applies not only to differences between intensities but also to the absolute limit of detection against darkness, because the absolute threshold is determined by the smallest signal which can be detected reliably against the random background of the neural noise, which is present in the visual brain even when no light enters the eye.

Implications of randomness

Randomness—events occurring unpredictably without apparent cause—has been seized on to rescue the nervous system (or rather our view of it) from being a machine without volition or free will. But this leads us to puzzling questions: How can we be responsible for actions, whether the nervous system is precisely *determined*, or if it is partly *random*? It is sometimes argued that randomness allows free will, but how can we be responsible, or take credit, for actions set off by chance events? Here neurology meets philosophy and ethics. The debate continues.

What can nerves signal?—only specific qualities

The founding father of modern physiology, Johannes Muller (1801–58), propounded a fundamental law of nerves: that a nerve fibre can only signal one kind of quality. Known as the Law of Specific Energies (or better, Law of Qualities), it is the key to how sensory systems, such as eyes and ears, are designed through Natural Selection to respond to complex characteristics of objects, including movements and colours.

All sensory nerves are essentially the same, signalling with just the same action potentials—yet they may give very different sensations—of brightness, pain, weight, tickle, sound, . . . and far more. What we experience depends on which region of brain a nerve is connected to. So if optic nerves were connected to the auditory cortex we would hear lights! The nerves signalling colours are identical, so how can we see different colours? This works because long, medium, and short wavelengths of light stimulate different nerves in the brain. How this works is described in Chapter 7 (shown in Figure 7.3).

For seeing movement, there are circuits which convert complex movements at the eyes to the usual simple electrical pulses of action potentials, which—coming from the movement-detecting circuits—can signal directions and speeds of motion of objects or of the observer. It may need higher and sometimes cognitive (knowledge-based) brain processes to make sense of movement, or any other sensory signals. Let's look at this in more detail: first for movement, then for colour.

6

Seeing movement

Detection of movement is essential for survival of all but the very simplest creatures. Moving objects are likely to be dangerous prey, or potential food, or a mate. They generally demand action of some kind, while stationary objects may be ignored with safety. Indeed, it is only eyes quite high up the evolutionary scale that produce signals in the absence of movement.

We can experience something of the long evolutionary development of vision, from the simplest creatures to ourselves, embalmed in the human eye.

The edges of our retinas are sensitive only to movement. You can see this by getting someone to wave an object around at the side of your visual field where only the edge of the retina is stimulated. Movement is seen, but it is impossible to identify the object, and there is no colour. When movement stops the object becomes invisible. This is as close as we can come to experiencing primitive vision. The extreme edge of the retina is even more primitive: when it is stimulated by movement we experience nothing; but a reflex is initiated, rotating the eye to bring the moving object into central vision, bringing our highly developed foveal region into play for identifying the object.

Eyes like ours which move in the head give information of movement into two distinct ways. When the eye remains stationary, the image of a moving object will run sequentially across many receptors, giving motion signals from the retinas. But when the eyes follow a moving object, the images remain essentially stationary on the retinas—so they cannot signal movement though we still see movement. This is so even when there is no background to sweep across the retina as the eyes move. This effect can be demonstrated with a simple experiment. Ask someone to wave a dim pencil torch slowly side to side in a dark room, and follow it with your eyes. You will see the movement of the light even though there is no image moving across your retinas. So evidently the rotation of the eyes in the head can give perception of object motion in the absence of movement signals from the retinas. This works because rotations of the eyes are signalled to the brain—in an unexpected way, as we will see.

There are, then, two essentially different visual systems for detecting movement: the *image–retina* system, and the *eye–head* system (Figure 6.1).

Image–retina movement

Recording electrical activity from the eyes reveals that there are various kinds of receptors, almost all signalling only changes of illumination; very few fire from continuous steady light. Some receptors signal when a light is switched *on*, others when it is switched *off*, while others again signal when it is switched *on* or *off*. These are named 'on', 'off', and 'on–off' receptors, respectively. It seems that those receptors responding only to changes of illumination are responsible for signalling movement, and that all eyes are primarily detectors of movement.

By placing electrodes in the retinas of frogs' eyes, it has been found that analysis of the receptor activity for signalling movement takes place in the frog's retina. Horace Barlow, at Cambridge, discovered retinal 'bug detectors' in the frog's retina, which elicit reflex tongue fly-catching, faster than signals processed by the brain. Feature detectors were described in a paper charmingly called 'What the frog's eye tells the frog's brain', by J. Y. Lettvin and colleagues in 1959, at the Massachusetts Institute of Technology. They found:

(1) fibres responding only to sharply defined boundaries;
(2) fibres responding only to changes in the distribution of light;
(3) fibres responding only to a general dimming of illumination, such as might be caused by the shadow of a bird of prey.

The physiological discovery that movement is specially coded as neural signals in the retina, or in the visual projection areas of the brain, is important in many respects. In particular, it shows that speed is signalled without involving a clock, or estimate of time. It is, however, sometimes assumed that neural systems giving perception of velocity must refer to an internal biological clock of some kind. This assumption arises from the fact that velocity is defined in physics by the time taken for something to travel a given distance, ($v = d/t$). But the speedometer of a car has no clock associated with it. A clock is,

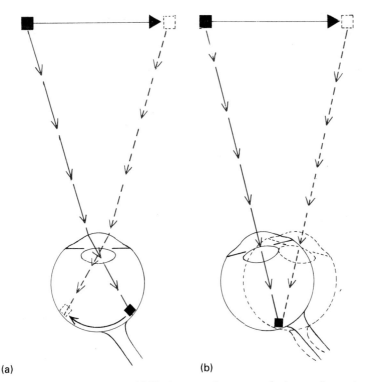

(a)   (b)

6.1 Two movement systems. (a) The image–retina system: the image of a moving object runs along the retina when the eyes are held still, giving information on movement through sequential firing of the receptors in its path. (b) The eye/head system: when the eye follows a moving object, the image remains stationary upon the retina, but we still see the movement. It is signalled from the commands to move the eyes. The two systems can sometimes disagree, to give paradoxical illusions of movement.

indeed, needed for calibrating such an instrument in the first place, but once calibrated, it will give velocity measures without the use of a clock. The same is evidently true of the eyes. The image running across the retina sequentially fires receptors in its path, and the faster the image travels (up to a limit) the greater the velocity signal this will give—with no need for a clock. Although analogies with other velocity detectors, such as speedometers, show that velocity could be perceived without reference to a clock, they do not tell us precisely how the visual movement system works. A detailed model has been suggested for the compound eye of flies, and is the basis for a system in aircraft that detects drift due to wind blowing them off course. This movement detector was developed by biological evolution millions of years ago, and has now been discovered by applying electronics, to be useful for technology. This is a nice example of backwards and forwards bio-engineering. The human eye has different mechanisms for very slow and for fast movement, and different mechanisms for short- and long-range movements of textures, or random dots. These make engineering sense.

It is a strange thought that we are engineering devices; the deep question is whether this extends right into brain and mind. The Oxford philosopher, Gilbert Ryle, famously spoke of the 'Ghost in the machine' Should we turn this around—to question the machine in the ghost?

Eye–head movement

The neural system giving perception of movement by shifting of images across the retina must be very different from the system for detecting movement by rotation of the eyes in the head. Somehow, the fact that the eye is being moved is signalled to the brain and used to indicate movements of external objects. This is demonstrated with the moving torch experiment described above, for in that situation there is essentially no movement across the retina and yet the movement is seen (Figure 6.1(b)).

The most obvious kind of eye–head signal would be from the eye muscles; when they are stretched signals would be fed back to the brain, indicating movement of the eyes and so of objects followed by the eyes. This would be the engineer's normal solution, but is it Nature's? We find the answer when we look at what may seem a different question:

Why does the world remain stable when we move our eyes?

The retinal images run across the receptors whenever we move our eyes, and yet we do not experience movement: the world does not usually spin round when we move our eyes. This does happen when a cine or video camera is panned round.

We have seen that there are two neural systems for signalling movement: the image–retina and the eye–head systems. It seems that during normal eye movements these signals cancel each other out, to give stability to the visual world. This was discussed by the physiologist who did most to unravel spinal reflexes, Sir Charles Sherrington (1857–1952), and by Helmholtz (Figure 6.2); but they had different ideas as to how it comes about—especially about how the eye–head system conveys its information of movement. Sherrington's theory is known as the *inflow theory* and Helmholtz's the *outflow theory* (Figure 6.3). Sherrington thought that signals from the eye muscles were fed back into the brain when the eye moves, to cancel the movement signals from the retina. But neural signals from the eye muscles would take longer to arrive than signals from the retinas, so we might expect a sickening jolt, just after movements of the eyes, before the inflow signals reach the brain to cancel the image–retina signals.

Helmholtz had a very different idea. He thought that retinal movement signals are cancelled not by 'inflow' signals from the eye muscles, but by central 'outflow' signals from the brain, commanding the eyes to move.

The issue can be decided by very simple experiments, which you can try yourself. Try pushing one eye gently with a finger, having closed the other eye by holding a hand over it. When the eye is rotated passively in this way, the world will be seen to swing round, in the opposite direction to the movements of the eye. Evidently, stability does not hold for *passive* eye movements, though it does for the normal *voluntary* eye movements. Since the world swings round *against* the direction of the passive eye movements, it is evident that the image–retina system still works; it is the eye–head system which is not operating. Why should the eye–head system work only for voluntary, not for passive, eye movements? Sherrington thought that the eye muscles send signals from stretch receptors, in the same way as stretch receptors are known to give feedback signals from the muscles which move the limbs. But although the six eye muscles do have stretch receptors, the eye–head system does not work this way. As we have just said, the stretch receptors would deliver signals too late to cancel the image–retina signals, without a sickening delay, and they are not accurate.

We may stop all retinal movement signals and see what happens during passive or active movements of the eyes. This is easily done by staring at a bright light (or a photographic flash) to get an after-image—like a photograph stuck on the retina. This will move precisely with the eye, so it will not give any image–retina movement signals. If you observe the after-image in darkness (to avoid a background) you will find that, if the eyes are pushed with the finger to move passively, the after-image does not move. This is very strong

6.2 Hermann von Helmholtz (1821–94)—physiologist, physicist, psychologist, philosopher—is the greatest figure in the experimental study of vision. His *Physiological optics* is the bedrock of the subject.

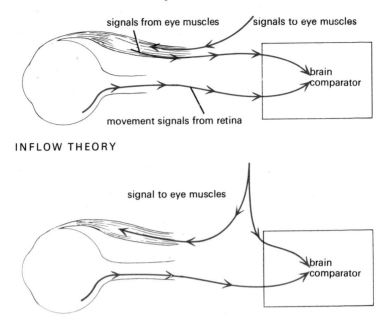

signals from eye muscles   signals to eye muscles

brain
comparator

movement signals from retina

INFLOW THEORY

signal to eye muscles

brain
comparator

OUTFLOW THEORY

6.3 Why does the world remain stable when we move our eyes? The inflow theory suggests that movement signals from the retinas (image–retina system) are cancelled by afferent signals from stretch receptors of the eye muscles. The *outflow theory* suggests that the retinal movement signals are cancelled by the efferent command signals to move the eyes (eye–head system), through an internal monitoring loop. The evidence favours the outflow theory.

evidence against the inflow theory. There will be signals from stretch receptors, but they do not produce sensations of motion.

However, if you move your eyes voluntarily, with the after-image viewed in darkness, you will see the after-image moving with the eyes. Helmholtz explained this by supposing that it is not signals from the eye muscles which are involved, but *commands* to move the eyes. This suggests that the command signals to move the eyes are monitored by an internal loop (Figure 6.3), to cancel the image–retina signals. When these are absent (as when an after-image is viewed in darkness) the world swings round with the eyes because the command signals are not cancelled by retinal motion signals. Passive movements of the eye give no movement of the after-image because in these circumstances neither system gives a movement signal.

In clinical cases, where something is wrong with patients eye muscles or their nerve supply, the world swings round for them when they try to move their eyes—moving in the direction their eyes should have moved. This also occurs if the muscles are prevented from functioning by curare, the South American arrow poison which blocks neural signals to muscles. The Austrian physicist Ernst Mach (1838–1916), bunged up his eyes with putty, so they could not move, to get the same result. (Not to be repeated!)

So the eye–head system does not work by detecting actual movement of the eyes, but from commands to move them. It works even when the eyes do not obey the commands. It is surprising that command signals can give rise to perception of movement; we usually think of movement perception as always coming from the eyes, not from a source deep in the brain controlling them.

Why should such a peculiar system have evolved? It is even more surprising when we find that there are in fact stretch receptors in the eye muscles. It seems that an inflow or feedback system would be too slow: by the time a feedback signal got back to the brain to cancel the retinal movement signal, it would be too late. In fact, because the signals from the retina have further to travel to reach the brain, the command signal could arrive too soon. Evidently it is delayed to match the small inevitable retinal delay (the retinal action time) so it does not arrive before the motion signal from the retina. We can literally see this in the movement of the after-image with voluntary eye movements. Whenever the eye is moved, the after-image takes a little time to catch up. Can one imagine a more beautiful system?

The case of the wandering light

You might like to try the following experiment. The apparatus is a small dim light (such as an LED) placed at the far end of a completely dark room. If you look at the light for more than a few seconds, it will wander around, in a curious erratic manner, sometimes swooping in one direction, and sometimes oscillating gently to and fro. Its movement may seem paradoxical: it may appear to move, yet not change its position.

The wandering light, is known as the *autokinetic phenomenon*. It has been the subject of a great deal of discussion and experimental work. A dozen theories have been advanced to explain it, and it has been used as an index of suggestibility and group interaction by social psychologists. The explanations are extraordinarily diverse.

One theory is that small particles drifting about in the aqueous humour, in the front chamber of the eye, may be dimly seen under these conditions. This theory then proposes that the spot of light, rather than the particles, seems to move just as the moon seems to scud through the night sky when clouds are driven by the wind. This 'induced movement' will be discussed later (pages 113–14). There is, however, plenty of evidence that this is not responsible for the auto-kinetic effect. The movements occur in directions unrelated to the drift of the eye's particles (when these are made clearly visible, with oblique lighting) and in any case the particles are not generally visible.

A very different theory, held by some ophthalmologists, is that the eyes cannot maintain their fixation accurately on a spot of light viewed in darkness, and the drifting of the eyes causes the image of the spot of light to wander over the retina, so that the light seems to move. This theory was all but disproved in 1928 by Guilford and Dallenbach, who photographed the eyes while subjects observed a spot of light and reported movements they saw. When the reported movements of the spot were compared with the photographic records of the eye movements, no relationship was found between them. In addition, the eye movements under these conditions were extremely small, while the autokinetic movements can be huge. This experiment seems to have gone largely unnoticed.

Attempts to explain the wandering of the light in the dark generally suppose that something is actually moving—particles in the aqueous humour, the eyes, or some sort of reference frame in the brain. This last suggestion formed an important part of the Gestalt psychologists' theory of perception, and they attached great weight to the wander-ing-light effect. Kurt Koffka, in his celebrated *Principles of Gestalt psychology* of 1935, says of it:

> The autokinetic movements are the most impressive demonstration of the existence and functional effectiveness of the general spatial framework, but the operation of this framework pervades our whole experience.

Is the argument sound? I believe it contains an important fallacy. What is true for the observed world of objects, does not necessarily hold for errors of observation, or illusions. It is important to appreci-ate the difference. Any sense organ can give false information: pressure on the eye makes us see light in darkness; electrical stimula-tion of any sensory nerve endings will produce the experience nor-mally given by that sense. (See page 97).

If movement is falsely signalled by some kind of neural distur-bance, we should expect to experience movement though nothing is actually moving. This is familiar from man-made instruments: the speedometer of a car may become stuck at a reading of say 50 m per hour and will indicate this speed though the car is stationary, locked

up in its garage. Similarly, pressure on the eyes makes us see light
when there is no light.

It is true that all real movement of objects in the world is relative,
and we can only speak of or measure movement of objects by reference to other objects. This, indeed, is the basis of Einstein's theory of
special relativity. This point was made two hundred years earlier,
by Bishop George Berkeley, when he challenged a point in Newton's
Principia, of 1687:

If every place is relative, then every motion is relative . . . Motion cannot be
understood without a determination of its direction which in its turn cannot
be understood except in relation to our or some other body. Up, Down, Right,
Left all directions and places are based on some relation and it is necessary to
suppose another body distinct from the moving one . . . so that motion is
relative in its nature.

Therefore, if we suppose that everything is annihilated except one globe, it
would be impossible to imagine any movement of that globe.

It has been assumed by many writers on perception, that if nothing
is moving—not the eyes, particles in the eyes, nor anything else—it
would be impossible to experience even *illusions* of movement, for
example of the spot of light in darkness. The wandering light has been
taken to represent the same situation as Berkeley's globe, when everything except it is annihilated; but it is actually very different.

The error lies in supposing that *false* estimates of movement—
illusions of movement—require something moving relative to something else. But they can result simply from a disturbance or a loss of
calibration of the measuring instrument, whether it be a speedometer
or the eye. We may now seek the kind of disturbance or loss of
calibration of the visual system responsible for the wandering light.
To do this, we will produce large systematic illusory movements of
the spot of light by deliberately upsetting the eye–head system.

If a person looks hard in any direction for several seconds, and then
returns the eyes to their normal central position, while a small dim
light is viewed in darkness, the light will usually be seen to speed in
the direction in which the eyes were held. It sometimes moves in the
opposite direction, but seldom in any other. The illusory movement
may continue for several minutes, when the eye muscles are
asymmetrically fatigued (Figure 6.4). Now fatigue of the eye muscles
requires abnormal command signals to hold the eyes' fixation on the
stationary light; but these are the same as the signals which normally
move the eyes, when they follow a moving object. We thus see movement when some of the muscles are fatigued, although neither the

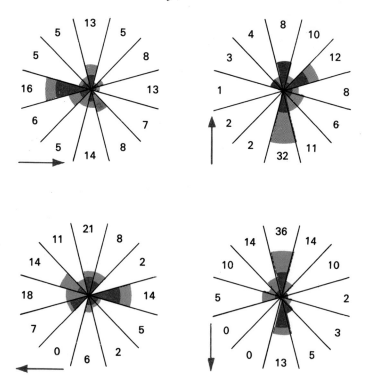

6.4 These 'clock histograms' show how a small dim light viewed in darkness appears to move after straining the eyes in one of four different directions for 30 seconds each time. The arrows show the direction of strain; the dark tinted areas show illusory movement during the next 30 seconds, while the light tinted areas show the following 30 seconds. The numbers give the duration in seconds of illusory movement in each direction during the first two minutes after strain.

eyes nor the image on the retinas are moving. The usual wandering movements of the autokinetic effect seem to be due to command signals maintaining fixation in spite of slight spontaneous fluctuations in the efficiency of the muscles, which tend to make the eyes wander. It is not the eyes moving, but the correcting signals applied to *prevent* them moving which cause the light apparently to move in the dark.

We may now ask: if the correction signals move the spot of light in the dark, why don't they cause instability of the whole visual world, in normal conditions? Why is the world generally stable? There is no certain answer to this question. It may be that in the presence of large fields of view, the aberrant signals are ignored, because the brain assumes that large objects are stable, unless there is clear evidence to the contrary. This is borne out by the effect of 'induced movement' (page 113). But first we should recognize that sometimes the world does swing round:

The case of the wandering world

The world swings around when we are very fatigued, or suffering the less pleasant effects of alcohol. This was described by the Irish wit Richard Brinsley Sheridan. After a party, friends led him to the front door of his house in Berkeley Square and left him. Looking back, they saw him still standing in the same position. 'Why don't you go in?' they shouted. 'I'm waiting until my door goes by again—then I'll jump through!' replied Sheridan.

Just how this ties up with the wandering light is not entirely clear. It may be that the eye movement command system is upset; or it may be that alcohol serves to reduce the significance of the external world, so that error signals normally disregarded are accepted. Just as we can become possessed by fantasies and irrational fears when tired or drunk, so might we become dominated by small errors in the nervous system, which are generally rejected as insignificant. If this is so, one might expect schizophrenics to suffer from instability of their visual world, but I know of no evidence for this.

The waterfall effect

We have found that the illusory movements of the spot of light viewed in darkness are due to small disturbances of the eye–head system. We might now expect to find illusions of movement due to disturbances of the image–retina system—and indeed we do. Various regions of the visual field may appear to move in different directions, and at different rates. The illusory effects are bizarre, and sometimes paradoxical, for it is possible to see movement without change of position.

After-effects from a turntable

A rotary after-effect is produced by looking steadily for say half a minute at the central pivot of a rotating record player. If the turntable is stopped suddenly it will seem to continue rotating—but in the opposite direction.

Rotating a spiral (Figure 6.5) produces illusory expansion or contraction. This transfers to another object, such as someone's face. As different regions are affected differently (and can be rotations, and changes of size) the effect must be retinal—in the image–retina system.

6.5 When the spiral is slowly rotated it is seen to shrink, or expand, depending on the direction of rotation. When stopped it seems to do the opposite of the inducing stimulus. This illusory shrinking or expanding is paradoxical, as the spiral does not appear to change in size.

This best known image–retina disturbance was described by Aristotle. Just after watching a waterfall, he saw the bank of the river apparently moving. So it is known as the waterfall effect though it is produced by any continuous motion, including expansion or contraction. The after-effect is always in the opposite direction to the 'inducing stimulus.

This can also be shown by following a moving belt of stripes with the eyes, returning the eyes rapidly to the beginning (with the lights switched off), following the movement again, and repeating this several times (Figure 6.6). In this way movement in one direction is experienced using the eye–head, but not the image–retina system.

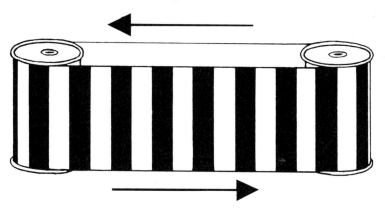

6.6 The waterfall effect. The stripes can be seen moving either by holding the eyes still (with a stationary fixation point) or by following the movement with the eyes. When the eyes are stationary, the image–retina system is stimulated. When they are following, the eye–head system signals movement—but not from the retina. Is there an after-effect from the eye–head system? It is certainly much less—probably absent.

When the belt is stopped after prolonged viewing with the following eye movements, there is no noticeable after-effect. So we may attribute the waterfall effect, at least almost entirely, to adaptation of the image–retina system. Only this could give illusory expansion or contraction.

Just why adaptation to retinal motion occurs is an interesting question. We have seen from the work of Hubel and Wiesel that movement is represented in separate neural channels, and that different channels indicate different directions of movement (Figure 4.8). It is reasonable to assume that some of these channels become 'fatigued' with prolonged stimulation, and this unbalances the system, giving illusory movement in the opposite direction. But it may have a use— for re-calibrating the system.

It remains a problem as to whether the adaptation takes place in the retina or in the brain. The retina may seem too simple for such a complex effect, but it is very difficult to rule out retinal adaptation as at least a part-cause. One might think (and several experimentalists who ought to have known better have thought) that the issue could be decided by looking at the moving stimulus with one eye, while holding the other closed, then noting whether the after-effect occurs when viewing a stationary object with the previously unstimulated eye. The answer is that it does occur, at about half strength. But this does not show conclusively that the adaptation took place in the brain—for it is possible that the previously stimulated eye goes on sending up movement signals after it is shut, and that these are 'projected' into the field of the unstimulated eye. This is perfectly possible, as one does not know which eye is active, although one tends to think, often wrongly, that whichever eye is open is doing the seeing. (To avoid this problem 'pressure blindness' has to be employed to block signals from the stimulated eye—but pressing the eye to impede blood flow to the retina is extremely dangerous; definitively not to be tried).

As we have said, if the after-effect from the rotating spiral is examined carefully, it is seen that the illusory movement is paradoxical—for example, the spiral expands without getting larger (Figure 6.5). So it is changing yet not changing. This is impossible for real objects, but we must remember that what is true for real objects may not hold for perception. We can even experience things that are logically impossible. In this case of the after-effect of movement, of expansion with no change in size, we may suppose that this comes about because *velocity* and *positions* are signalled by separate neural channels. When one of these is upset, it will disagree with signals from the other channel. So the brain is in the position of a trial judge getting conflicting evidence from witnesses. When incompatible stories are accepted, the judge is stuck with a paradox. As the perceptual system has many channels and many sources of information, the brain must serve as judge—sometimes rejecting channels, or sources of information, sometimes accepting incompatible signals, or conflicting information. Then we experience a paradox. We should not be too surprised that illusions and hallucinations can be impossible to describe in terms of normal experience.

Relativity of movement

So far, we have considered the basic mechanisms for perceiving movement, either by stimulation of the retina by moving images or by the eye following moving objects. There is, however, far more to the perception of movement. Whenever there is motion the brain has to decide what is moving and what is stationary. Although, as we have seen, it is fallacious to suppose that illusory movement necessarily involves any *actual* movement, it remains true that all real movement is relative, and so a decision is always required about what is moving. An obvious example occurs whenever we change our position by walking, driving, or flying. We generally know whether the movement is due to our own motion among surrounding objects, rather than due to their movement; but this involves a decision. Sometimes the decision is made wrongly, giving errors which can be particularly dangerous, because perception of movement is of such prime importance for survival. This is as true in the case of humans living in an advanced civilization, as ever it was in the primitive state in which eye and brain evolved.

Most perceptual research has been undertaken with the observer stationary, often looking into apparatus showing flashing lights or pictures. But real-life perception occurs during free movement of the observer, in a world where some of the surrounding objects are also in motion, such as when driving a car in traffic. There are severe technical problems for investigating such real-life situations but the attempt is worth while, as the results are important for driving, flying, and also space flight. Astronauts are trained on highly realistic simulators to make judgements of speed, and size, and distance. The same is true for pilots. Visual simulators are important for flying and space training, and the simpler aspects of driving; but simulators never copy reality quite perfectly, so skills learned from them are seldom quite appropriate for the real thing. Discrepancies can show up at awkward moments as bad habits—'negative transfer', from the simulator.

As we have seen, there is always a decision involved as to what is moving. If the observer is walking, or running, there is seldom a problem; for he or she has a lot of information from the limbs, signalling movement in relation to the ground. But when carried along in a car, or an aircraft, the situation is very different—the feet are off the ground, and the only source of information is the eyes. This is so except during acceleration (or deceleration, or moving in a curved path) when the balance organs of the middle ear give some, though often misleading information.

Induced movement

The phenomenon known as induced movement was investigated by the Gestalt psychologist Karl Duncker. He devised demonstrations to show that, when there is relative motion, we generally see the largest object as stationary, the smaller as moving. Thus we tend to see the small moon as rushing across the sky, though it is the large clouds that are moving past. This is not easy to demonstrate in a room, but is just possible by projecting a small patch of light on a large movable screen (Figure 6.7). When the screen is moving, what is *seen* as moving is the small stationary patch of light.

It may be noted that actually there is some information available on what is moving—whether the eyes rotate. But the eye–head signal is

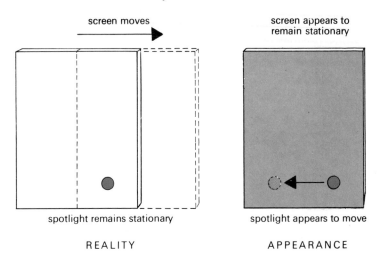

6.7 Induced movement. A small patch of light is projected on a large screen, which is moved. This makes the small stationary light appear to move in the opposite direction.

not always sufficient to decide the issue. This is relevant to the curious question of why the visual world does not swing round with eye movements, though it does when a cine camera is panned around a scene (page 101).

It seems clear that the brain bets on small objects moving, as generally large things (trees and houses) are stationary. When the bet is wrong, it can be disturbing and dangerous, as when driving a car: 'Is it my hand-brake that is off—or is that idiot over there running backwards?' The answer is important!

Motion and distance

When we move, say to the right, near objects move against our motion—to the left—and further objects seem to move with us, to the right. This is motion parallax, and it has a simple explanation. It is the same effect as the slightly different views of each eye giving stereoscopic vision, and indeed (as Brian Rogers at Oxford University has pointed out) stereo depth perception may have evolved from motion parallax—which itself is a valuable cue to depth. Although the geometry of dynamic parallax is simple, there are surprisingly subtle visual effects.

When we observe the moon while travelling in a car at night, we see it apparently moving along with us—rather slowly. At 50 km per hour, the moon may seem as moving at say 10 km per hour. We see it as moving more slowly than us yet keeping up—again a paradox.

The moon is so distant, we can regard it as at infinity. As the car moves along, the angle to the moon remains unchanged. So unlike near objects it does not change its position, though we are moving along under it. But perceptually it lies at a distance of only a few hundred metres. We know this from its apparent size. It subtends an angle of 0.5°, but looks the size of say an orange a few hundred metres away. The moon does not get left behind, because it is so distant that its angular position does not change—yet it appears to be not so far away. The only way the perceptual system can reconcile these, is to 'assume' that the moon is moving with the car. The apparent velocity of the moon is set by its apparent distance. If we change the apparent distance (by viewing it through converging prisms to change convergence of the eyes) then it seems to move at a different speed.

A particularly interesting, closely related effect, can be seen by projecting stereo pictures on a screen (with crossed polaroids, or red–green glasses). When the observer moves his or her head from side to side, the stereo scene seems to rotate, to follow the observer's every movement although it is fixed on the screen. Thus, a stereo projection of a corridor swings round, so the apparent front moves with the observer. This is opposite to the motion of a true scene— such as the real corridor. There is more: if the stereo pictures are set to coincide far away, at the end of the corridor, it will not rotate, but will shift equally along its entire length. If now, the stereo pictures are made to coincide near the middle of the corridor (so there are opposite disparities at its far and near ends), the corridor appears to rotate around the point of zero disparity, still *with* the observer's movements.

Stereo projection is revealing, because the scene lies flat on the screen, so there is no physical motion parallax even though it is seen in depth. Normally the world rotates around the point of fixation of the eyes, in the opposite direction to the observer's motion. (This is clearly seen while travelling in a train). But observing flat pictures in stereo depth, the opposite happens; they appear to rotate *with* the observer's movements, around the point of convergence of the eyes where disparity is zero—set by the separation of the stereo-pair on the screen. These effects are well worth trying out. As these illusory movements are in the opposite directions to normal motion parallax, evidently what we are seeing are *compensations* to normal motion parallax. It also works for ordinary pictures seen realistically in depth.

Movements of life

The Swedish psychologist Gunner Johansson has produced extremely striking demonstrations of how little information is needed for seeing moving humans, and animals. If small lights are placed at the joints of someone's arms and legs, he or she is effectively invisible in a dark room—until he or she moves (Figure 6.8). Immediately the lights are seen as a human figure, and can be identified as male or female, from the slight differences of movements. This does not work well for unfamiliar objects, and works best with living beings, especially humans.

The ability to recognize friends and foes has always been very important for survival, especially perhaps at night, and where the person or animal is partly hidden. So very little information is needed, when there are movement signals. The more probable the object, the less information is needed. This is a wonderfully powerful demonstration of the important perceptual principle that we can see more than meets the eye.

Movement in cinema and television

All the sensory systems can be fooled. Perhaps the most persistent fooling is by cinema and television. They present a series of still pictures, but what we see is continuous action. This relies on two distinct (though often confused) visual principles: *persistence of vision* and the *phi phenomenon*.

Persistence of vision

This is simply the inability of the retina to signal rapidly changing intensities. A light flashing at a rate greater than about 50 flashes per second appears steady; for bright lights this critical fusion frequency (CFF) may reach 100 or more flashes per second. Cinema presents 24 frames per second; but a three-bladed shutter is used to raise the flicker rate to 72 flashes per second—three for each picture. Television (British standard) presents 25 fresh pictures per second, each given twice, to raise the flicker rate to 50 per second. Television flicker is reduced by 'interlaced scanning', in which horizontal sections of the picture are built up by scanning in bands of lines, rather than continuously down the screen. Nevertheless, television flicker can be annoying, and is potentially dangerous to sufferers from epilepsy. Indeed, flicker is used to evoke symptoms for diagnostic purposes. Flicker can

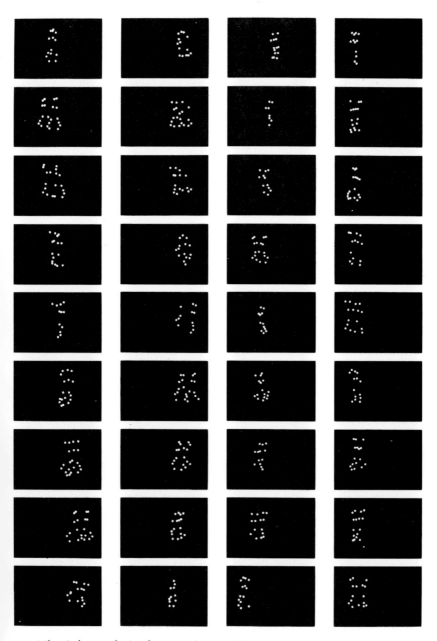

6.8 When lights attached to the joints of arms and legs move, two people are immediately seen. (After Johansson 1975, © 1975 Scientific American, Inc., all rights reserved.)

be a hazard when driving along a row of trees with a low sun. When landing a helicopter in tropical conditions, the rotor blades produce a violently flickering light which can be dangerous. The 'stroboscopes' used in discotheques can be similarly disturbing, and should probably be avoided; as also, of course, should the high levels of sound, capable of producing permanent hearing loss. The senses are too precious to be abused and damaged unnecessarily.

Low-frequency flicker produces very odd effects on normal observers, as well as on those with a tendency to epilepsy. At flash rates of five to ten per second, brilliant colours and moving and stationary shapes may be seen and can be extremely vivid. Their origin is obscure, but they probably arise from direct disturbance of the visual system of the brain, the massive repeated bursts of retinal activity overloading the system. The patterns are so varied it is difficult to deduce anything about the brain systems which have been disturbed. This can be an unpleasant experience, leading to headache and nausea.

Phi movement

The other basic visual principle upon which cinema depends is the apparent movement known as the phi phenomenon. It is usually studied in the laboratory with a very simple display, just two lights, switched so that just after one light has gone off the other comes on, in sequence. What is seen—when the distance between the lights and the time intervals between their flashes is about right—is a *single light*, apparently moving between the flashing lights. It was argued by the Gestalt psychologists that this apparent movement across the gap between the lights is due to an electrical charge, sweeping across the brain, to give sensation of motion and fill the gap. As the Gestalt psychologists, early this century, thought that the phi phenomenon demonstrated such a basic process, it was studied intensively. Probably all authorities would now, however, consider the Gestalt view of the matter fundamentally mistaken.

The simple notion that the image–retina system is tolerant of gaps explains phi movement. For vision needs tolerance, to cope with all manner of inadequacies. (The situation is like a key and a lock: the key does not have to be *exactly* a certain shape; some tolerance is essential, or any slight change in the lock or the key would stop it working.) This use of tolerance is a basic engineering principle.

Moving objects can momentarily disappear, as when a running animal for a moment is hidden behind a nearby tree; but it is useful for observers to see this as a continuous movement of the same object. The image–retina system tolerates gaps, provided the jumps in space and time are not too large. As a fortunate pay-off, this tolerance in space and time allows cinema and television to be economically possible.

Seeing causes

The importance of regular lawful relations for perception was studied experimentally by Albert Michotte, at Louvain in Belgium. For many years he investigated visual perception of causality, moving patches of colour with various velocities and time delays, using the apparatus shown in Figure 6.9. He arranged for one coloured patch to move towards, and sometimes to touch another which then moves off, generally after a small delay. With some combinations of velocity and delay, there was an irresistible impression that the first patch had struck the second—pushing it causally, as objects such as billiard balls

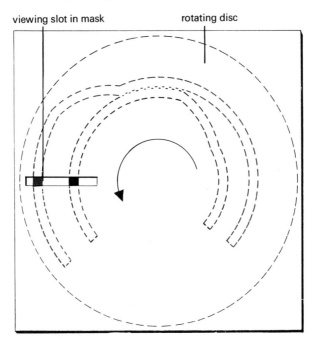

6.9 Michotte's apparatus for finding rules of perceived causality. A coloured patch moves towards another, which after a controlled delay moves away. This appears causal when delays and velocities approximate interactions of typical objects. (This can now be done far more simply with computer graphics.)

interact in the world of physics, though these were but coloured patches. Michotte thought that the seeing of causes is an innate response, but this is hard to prove, as it might be learned from years of seeing how objects interact.

Similarly, one sees causes in a cartoon film, even though the objects are abstract. This gets interesting when causal interactions are inappropriate to knowledge of the objects. Indeed, much of the humour of cartoons lies in disagreements between causal rules and object meanings. Here, again, we find a rich mixture of rules (syntax) and meaning (semantics) in perception. Cartoons are an ideal medium for playing upon and investigating primitives and sophistications of seeing.

Although Michotte was inclined to think that seeing causes is given innately because his observers had general agreement (though there were difficulties with differences in their verbal descriptions), we all encounter much the same kinds of objects, so shared experience might set up similar perceptions of cause. Observers' agreement over common situations does not distinguish innateness from learning from experience; but this is always a very difficult issue to resolve. Experiments with unnatural objects in computer games and virtual reality might give new evidence for settling this old controversy.

However this may be, causal relations in science are often very different from appearances—as 'arrows' of causal direction must be established by conceptual models. Here is an important difference between perceptual and conceptual hypotheses. Night follows day, but neither causes the other. We need a mental model of the Earth spinning round to 'see' the cause conceptually.

7

Seeing colours

The study of colour vision is an off-shoot from the main study of visual perception. It is almost certain that no mammals up to the primates possess colour vision. If any do, it is rudimentary. What makes this so strange is that many lower animals do possess excellent colour vision: it is highly developed in birds, fish, reptiles, and insects, such as bees and dragon-flies. We attach such importance to our perception of colour—it is central to visual aesthetics, and profoundly affects our emotional state—that it is difficult to imagine the grey (as we might think of it) world of other mammals, including pet cats and dogs.

The history of the investigation of colour vision is remarkable for its acrimony. The problems have aroused lifelong passions. The story is told of a meeting of 50 experts on colour vision, who defended 51 theories! Perhaps some theories never quite die; but it now seems clear that the first suggested is essentially correct.

The scientific study of colour vision starts with Newton's great work, the *Opticks* (1704). Surely this is the scientific book of its period most worth reading today. The highly imaginative and painstaking experiments were carried out in Newton's rooms (and others) in Trinity College, Cambridge, which still exist and are still lived in. It was there that his pet dog, Diamond, upset a lighted taper while Newton was in chapel, burning manuscript notes on his optical experiments. He delayed publishing *Opticks* until after the death of a rival genius, Robert Hooke, whose *Micrographia* (1665) is also well worth reading. The *Opticks* contains the famous 'Queries' which are Newton's most speculative thoughts on physics, and man's relation through vision with the universe.

Newton showed that white light is made up of all spectral colours. With the later development of the wave theory of light, it became clear that each colour corresponds to a given frequency. The essential problem for the eye, then, is how to get a different neural response for different frequencies. The problem is acute because the frequencies of radiation in the visible spectrum are so high—far higher than nerves can follow directly. In fact the highest number of impulses a nerve can transmit is under a thousand per second, while the frequency of light is a million million cycles per second. The problem is: How are the high frequencies of light represented by the slow-acting nervous system?

This problem was tackled for the first time by Thomas Young in 1801 (Figure 7.1). He suggested the theory, further developed by Helmholtz, which is now accepted, though it is not the whole story. Young's contribution was assessed 70 years later by the physicist James Clerk Maxwell in the following words:

It seems almost a truism to say that colour is a sensation; and yet Young, by honestly recognising this elementary truth, established the first consistent theory of colour. So far as I know, Thomas Young was the first who, starting from the well-known fact that there are three primary colours, sought for the explanation of this fact, not in the nature of light but in the constitution of man.

If there were receptors sensitive to every separable colour there would have to be at least 200 kinds of receptor. But this explanation must be impossible—for the very good reason that we can see almost as well in single-wavelength (monochromatic) coloured light, as in all-wavelengths white light. The effective density of the receptors cannot therefore be reduced very much in monochromatic light, and so there cannot be more than a very few kinds of colour-responsive receptors. Young saw this clearly. In 1801 he wrote:

Now, as it is almost impossible to conceive each sensitive point of the retina to contain an infinite number of particles, each capable of vibrating in perfect unison with every possible undulation, it becomes necessary to suppose the number limited, for instance, to the principal colours, red, yellow and blue.

Writing a little later, he stuck to the number of 'principal colours' as three, but changed them from red, yellow, and blue—to red, green, and violet—which we accept today.

We have now come to the hub of the problem: how can all the colours be represented by only a few kinds of receptor? Was Thomas Young right in supposing the number to be three? Can we discover the principal colours?

The possibility that the full gamut may be given by only a few principal colours is shown by a single and basic observation: colours can be mixed. This may seem obvious, but the eye behaves very differently in this respect from the ear. Two sounds cannot be mixed to give a different pure third sound; but two colours give a third

7.1 Thomas Young (1773–1829), after Lawrence. Young was the founder, with Helmholtz, of modern explanations of colour vision. A universal genius, Thomas Young made important contributions to science—especially diffraction and the wave theory of light—as well as to Egyptology where he had key ideas for translating the Rosetta stone.

colour, in which the constituents cannot be identified. Constituent sounds are heard as a chord, and can be separately identified, at any rate by musicians; but no training allows us to do the same for light.

We should be very clear, at this point, on just what we mean by mixing colours. When a painter mixes yellow and blue to produce green, he or she is not mixing lights; but is mixing the total spectrum of colours *minus the colours absorbed by the pigments*. This is so confusing that visual scientists generally forget about pigments and consider only mixing of coloured *lights*—which may be produced by filters, or prisms, or interference gratings (see Figure 7.2). Ignoring pigments is fine for thinking about colour vision and the eye, but rather misses the point for artists mixing paints.

When mixing lights, yellow is obtained by combining red with green lights. Young suggested that yellow is always seen by effective red–green mixture, there being no separate special receptor 'tuned' to be sensitive just to yellow light; but rather two sets of receptors, sensitive respectively to red and green, their combined activity giving the sensation yellow. What is combined are the nerve signals from the red and green receptors, whose responses overlap, so both are stimulated by what we see as yellow light. This is the presently accepted theory.

The fulcrum of controversies of colour theories has been the perception of yellow: is yellow seen by combined activity of red and green systems, or is it *primary*—as its simple perceptual quality might suggest? Although the pure appearance of yellow—it does not look like a mixture—has been raised against Thomas Young, this argument is not valid. The fact is that when a red and a green light are combined (by projecting the lights, overlapping on a screen) we do see yellow, and this sensation is indistinguishable from seeing one monochromatic light from the yellow region of the spectrum. In this instance at least, simplicity of sensation is no guide to simplicity of the underlying neural processes.

Young chose three principal colours, as he found that he could produce any colour visible in the spectrum (and white) by a mixture of three, but not less than three, overlapping lights set to appropriate intensities. He also found that the choice of suitable wavelengths was quite wide, and this was why it was so difficult to answer the question: what are the principal colours? If it were the case that only three particular colours would give by mixture the range of spectral hues, then we could say with some certainty that these would correspond to the basic colour systems of the eye; but there is no unique set of three wavelengths which will do the trick.

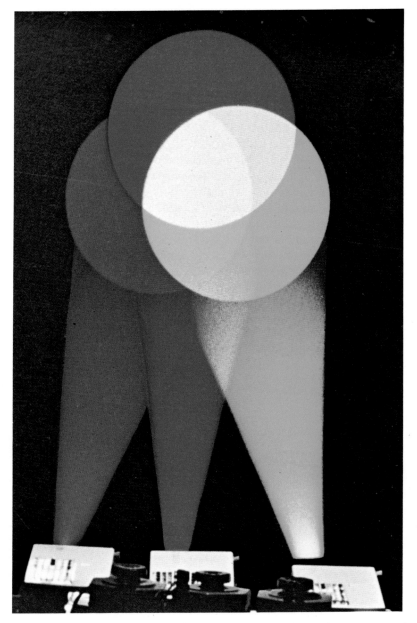

7.2 Thomas Young's experiment on colour mixture. By mixing three lights (not pigments) widely separated along the spectrum, Young showed that any spectral hue could be produced by adjusting the relative intensities. He could also make white, but not black, or any non-spectral colour such as brown or gold or silver.

Young's demonstration is beautiful. Figure 7.2 gives an idea of what it looks like.

The Young–Helmholtz theory is, then, that there are three colour-sensitive kinds of receptors (cones) which respond respectively to red, green, and blue (or violet), and that all colours are seen by a mixture of signals from the three systems. A great deal of work has gone into trying to isolate the basic response curves and this has proved surprisingly difficult (Figure 7.3).

We may now look at a further graph, basic for the understanding of colour vision, the so-called hue-discrimination curve (Figure 7.4). This compares wavelength of light with the smallest difference which produces a difference in observed colour, or hue. If we look at Figure 7.4, we see that hue should change very little at the ends of the spectrum when wavelength is varied, for the only effect is a gradual increase in the activity of the red or the blue systems. So there should simply be a change of brightness, with little or no change of colour. This is what happens. In the middle of the spectrum, on the other hand, we should expect dramatic changes in colour, as the red system rapidly falls in sensitivity and the green rapidly rises. A small shift of wavelength should produce a large change in the relative activities of the red and

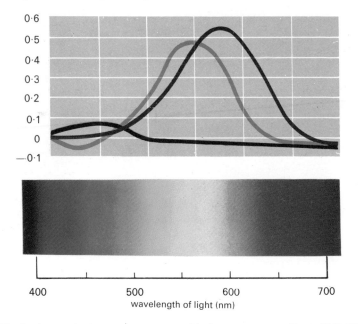

7.3 The fundamental colour response curves of the human eye, according to W. D. Wright. These represent the absorption curves of the three colour-sensitive cone pigments.

green systems, giving a marked change of hue. We should thus expect hue discrimination to be exceptionally good around yellow—and indeed this is so.

An alternative theory has been that there are four cone pigments—signalling 'opponent' colours. (One sees these as negative after-images after looking for several seconds at a bright coloured light.) Though no longer accepted, this theory has not entirely died, as there do seem to be opponent neural processes beyond the receptors, and those represent yellow.

For paintings and colour printing, pigments are used to remove—subtract—colours from white. The pigments (or filters used for most colour photography) are the 'complementaries' of Thomas Young's red, green, and blue primaries, which combined give *additive* colour mixture (Figure 7.2). Paints (and indeed the surfaces of all objects

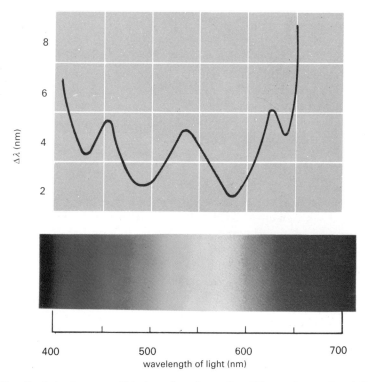

7.4 Hue discrimination curve. This shows how the smallest difference in wavelength ($\Delta\lambda$) varies with the wavelength (λ). It should be smallest—best hue discrimination—where the response curves (Figure 7.3) have their steepest opposite (one going up, the other down) slopes. This is roughly true.

which do not emit light) can only take away colours. We see what is left; so for instance green leaves absorb colours other than green.

One can think of how *subtractive* colours work in two ways: from the cone receptor response curves of Figure 7.3 and from the colour triangle of Figure 7.5. The colours used for subtraction lie across the colour triangle. In much the same way, complementary colours—as seen in after-images—occur across the triangle, because the unaffected colour receptors are relatively more active. Thus, following looking at a bright red light, a white light looks bluey-green.

So printers do not use red, green, and blue. Colour printing uses *cyan*, stimulating blue and green receptors; *magenta*, which stimulates red–blue, and *yellow* which stimulates red–green.

Printing the cyan subtracts red, the magenta subtracts green, and yellow subtracts blue. Subtraction of all colours gives black. Subtraction of none gives white, if we start with white paper. This is how the colour pictures in this book are printed. So for instance a red colour is produced by printing magenta and yellow inks (see Figure 7.6).

There is, however, a way of printing colours additively—by pointillism. As the French Impressionists discovered, particularly interesting effects can be produced with small, closely spaced dots of colour. When they are not seen separately but merge in the eye, pictures can be produced with wonderfully clear shimmering colour which may even outdo reality. This is also how colour television works.

Colour blindness

It is remarkable that the most common form of colour confusion—red confused with green—was not discovered before the late eighteenth century, when the chemist John Dalton found that he could not always distinguish between chemicals in glass bottles by colour though his friends could without difficulty. Why colour blindedness was missed for so long is presumably because we recognize objects by many criteria: grass is called 'green' as everyone calls it green, what-

---

7.5 Colour triangle. This represents 'colour space', arranged within the three primary colours of Thomas Young. White in the middle is given by equal activity of the three systems. Saturation increases outwards towards each primary. Mixture colours occur according to various contributions from the three systems. So this subjective colour map ties up with the primary systems, and how they overlap for different wavelengths of light. It also shows how several wavelengths produce mixture colours—as well as why we see negative after-images following exposure to bright coloured light. (Reproduced courtesy of Minolta (UK) Limited.)

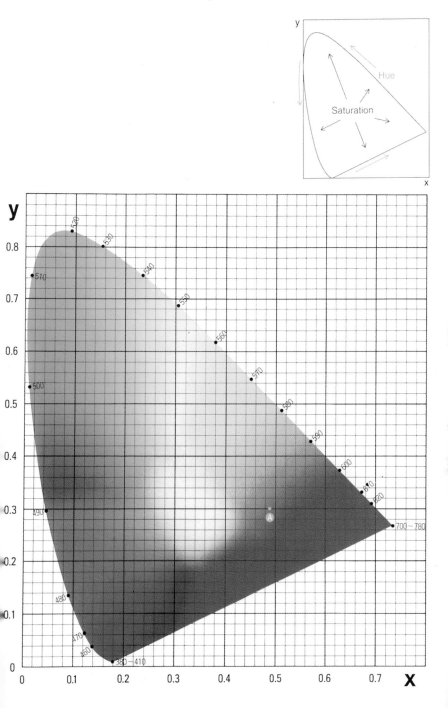

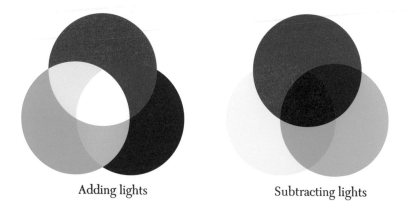

Adding lights   Subtracting lights

7.6 Adding coloured lights and colour printing by taking away from white.

ever they see. This is part of the puzzle: how do I know that *your* sensation of what I call 'green' is the same as mine? How can we know each other's sensations or 'qualia'? The answer is we don't and we can't. But we can recognize different abilities (or losses) of discriminations between for example colours.

For a chemist identifying substances in glass bottles there can be *only* the colours to distinguish them. This is the key to tests of individual colour vision. The most common confusion is between red and green; but for some people there are other confusions. Red–green confusion is surprisingly common: at least 10 per cent of men are markedly red–green deficient, though it is rare in women. (It is inherited by an X-chromosome-linked recessive gene). Much less common, for both sexes, is green–blue confusion.

Colour blindness is classified into three main types, based on the three receptor systems. They used to be called simply, red–green and green–blue blindness (or 'anomaly') but colour names are now generally avoided. Some people are completely lacking in one of the three cone systems: called *protanopes, deuteranopes,* or *tritanopes* (after the first, second, and third colour-receptor systems), but this does not clarify the situation very much. The essential feature is that these people need only two coloured lights to match all the spectral colours they can see. Thus Young's colour mixture result applies only to most people—not to extreme cases of colour blindness. It is more common to find, not a complete absence of a colour system, but rather a reduced sensitivity to some colours. These are classed as: *protanopia, deuteranopia,* and *tritanopia.* The last, tritanopia, is extremely rare. People with these deficiencies are (or should be) described as having anomalous colour vision. This means that, although they require mixtures of three coloured lights to make their spectral colours, they use different proportions, or relative intensities, from the normal.

The proportions of red and green light a person needs to match a monochromatic yellow is the most important measure of colour anomaly. It was discovered by the physicist Lord Rayleigh in 1881, that people who confuse red with green require a greater intensity, either of red or green, to match monochromatic yellow, (as from a low pressure sodium lamp). Special instruments (anomaloscopes) are made for testing the Rayleigh equation, which provide a mono-chromatic yellow field and a red-plus-green mixture field. The relative intensities of the red and green in the mixture field can be varied, until the mixture gives the same colour, to the observer being tested, as the monochromatic yellow field. The proportions are read off a scale which indicates the degree of protanopia or deuteranopia. The instru-ment is called an anomaloscope.

We have said that as yellow seems a pure colour it has been thought that there must be special yellow receptors tuned to its wavelength. But it can be shown quite simply with the anomaloscope that yellow—including monochromatic yellow—is always seen by mixture of the red and green receptors (see Figure 7.7).

If there were special yellow-sensitive receptors the match would be lost when the observer was adapted to red or green light, because the receptor tuned to the yellow wavelength would not be shifted by adaptation and would still signal yellow. If yellow is seen by the red and green receptor systems, however, it should be shifted when the sensitivity of the red or the green systems is affected (reduced) by the adapting light. This is equivalent to changing their relative intensities in the mixture field, and is indeed what happens. Whenever the match holds after adaptation, there cannot be different receptors signalling the fields. So there is no special yellow receptor. The experiment may be repeated for other colours, with a similar result, showing that no colour as seen has a special system: all are seen by mixture, except extreme red and extreme blue which do not change colour with adaptation.

The same result is obtained for anomalous observers. Their initial setting is of course different—or the instrument wouldn't detect an abnormality—but it remains unaffected by the colour adaptation.

This takes us to a curious conclusion. If the anomaloscope cannot distinguish between the normal eye with and without colour adaptation (which enormously change how the fields in the instrument appear), how does it measure colour anomaly? The anomalous eye must be quite different from the colour-adapted eye. So anomaly cannot be

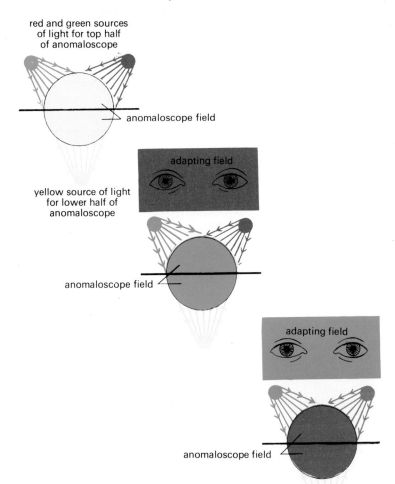

7.7 Disproving special receptors for yellow.

Observers adjust an anomaloscope so that they see an identical (normally yellow) colour in the mixture and in the monochromatic fields. They then look into a bright red light, to adapt the eye to red. While the retina is adapted to red, they look back into the anomaloscope and are asked to judge whether the two fields still look the same colour. Observers will now see both fields as green and they will be the *same* green. The match is not upset by the adaptation to red: so the observers will not require a different proportion of red plus green light in the mixture field to match the monochromatic yellow. It would therefore be impossible to tell from the setting of the anomaloscope that the eye has been adapted to red, though what the observer sees is quite different—a vivid green, instead of yellow in both fields. The same is true for adaptation to green light: both fields will now look the same red. In brief: the match still holds with adaptation to red or green light, though the seen colours are greatly changed by the adaptation. So both fields are seen by neural mixtures, though one is monochromatic.

simply reduction in sensitivity of the red or green systems. This is exactly what colour anomaly is often supposed to be, but this must be wrong. The reason for anomaly is not always clear but there may be many causes. It is not due to a straightforward shortage of photopigment, or the anomaloscope would not work, but some kinds of anomaly are due to spectral shift of response curves. Others might be due to neural 'short circuiting' of colour-receptor systems, so that the red and green systems signal more or less as though there were but a single system.

There is more to colour than meets the eye

We shall pass over the later, though now historical, acrimonious debates on whether there are three, four, or seven colour systems, and accept Young's notion that all colours are due to mixtures of three colours. But there is more to colour vision than is revealed by experiments with coloured lights. For we tend to see the colours of surfaces of objects as much the same in spite of large changes of the colour of the light illuminating the object. This is known as colour constancy. It implies that where seeing an object is concerned, we do not see colours simply according to wavelengths of light. It also implies that displays designed to be simple for quantitative measurements in the laboratory can miss essential features of perception.

A jolt was given to the more complacent, in the 1960s, by the American inventive genius Edwin Land. Apart from inventing Polaroid when still a research student, and later developing the instant camera, he showed, with elegant demonstrations, that what is true for colour mixture of simple patches of light is not the complete story for all perception of colours. Odd things happen when the patches are more complicated, and when they represent objects.

What Edwin Land did was to repeat Young's colour mixture experiment using not simple patches of light but photographic transparencies of patterns or scenes. We may indeed think of all (projection) colour photography as essentially Young's experiment put to work; for colour films provide only three colours. The same is true of colour television: there are only three colours of phosphors on the screen. But Land reduced the given colours to two, and found that a surprising wealth of colour is seen from only two wavelengths, when there are patterns or pictures.

The technique is to take two black and white photographic negatives of the same scene—each through a different colour filter. The negatives are converted into positive transparencies, and projected through their original filters, to give two superimposed pictures on the screen. Almost every colour will be seen.

Surprisingly good results are obtained with a red filter on one projector and no filter on the other. From Young's original experiment we would expect nothing but pink; but instead, we get green and other colours not physically present. This kind of result might, however, have been anticipated from two well-known facts. Some early colour films used only two colours—later it was realized how surprisingly good these could be. Second, although Thomas Young found that the *spectral* hues and white could be produced by mixtures of three lights, it is not possible to produce *every* colour that can be seen by using the three lights, whatever their proportions. For example brown cannot be produced, and neither can the metallic colours gold and silver. So there is something odd about the successful use of *three* colours, let alone two.

Consider an ordinary photographic colour transparency, projected on to a screen. This gives us pretty well all the colours we ever see; yet it consists of only the three lights of Young's experiment. The colour film is no more than a complex spatial arrangement of three coloured filters, yet this gives us brown and the metallic and other colours Young was unable to produce with his three simple overlapping discs of colours (Figure 7.2).

This means that any simple account of colour vision is doomed to failure: colour depends not only on the stimulus wavelengths and intensities, but also on differences of intensity between small regions, and whether the patterns represent objects. One might think that this must require high-level processes in the brain; but amazingly, bees have been shown to see in a similar way! The physiologist Semir Zeki, working in London, has shown recently that in the visual area of the chimp brain, there are neurones responding to 'Land colours'. These are early in the processing of signals from the eyes (visual area V4), so this does not seem to be a high-level process. It does seem to be related to colour constancy, which has been shown for eyes and brains as relatively simple as those of bees.

The eye tends to accept as white not a particular mixture of colours, but rather the general illumination whatever this may be. Thus we see a car's headlamps as white while on a country drive; but in town where there are bright white lights for comparison, they look quite yellow. The same is so for candlelight compared with daylight. This means that the reference for what is taken as white can shift. Expectation, or knowledge, of the normal colour of objects is important:

oranges and lemons take on a richer colour when they are recognized; but this is certainly not the whole story. Land was careful to use objects whose colours could not have been known to the observers— objects such as reels of plastic-covered wire, and coloured wool. He still got dramatic results in his two-colour projections, though the colours were unknown.

There is a conflict between designing experiments simple enough for analysis and sufficiently complex to reveal the full richness of perceptual phenomena. So science is an art. Like the arts, it is not completely mastered.

8

Learning how to see

An ancient question in philosophy is: how do we come to know the world? Are we born with inherited, or innate, knowledge—or do we have to learn everything we know? Do newborn babies have to learn how to see?

Philosophers are divided into those often termed metaphysical, who hold that we do have some knowledge of the world apart from any sensory experience, and on the other hand empiricists claiming that all knowledge is derived from observations, experiments, and measurements. To the metaphysician, it seems clear that by thinking sufficiently hard and in the right kind of way it is possible to make 'contingent' discoveries, such as the number of planets, without having to look. To empiricists, perception is basic for all knowledge; though, it must be admitted, knowledge and assumptions can affect perception.

The psychologist is concerned with development rather than with what is inherited at birth. But how much of a child's development is pre-programmed from inheritance? The present view is that the individual behaviour of the child is very important; so development is complicated results of interactions with the world of objects, in social situations, very much depending on the child's initiatives and available opportunities for play and discovery. Child's play is not unlike the exploratory play of scientists in their nurseries—their laboratories.

Much innate, immediately available knowledge is found in animals, most dramatically 'releasers' such as fledglings cowering from the profile-shapes of dangerous birds of prey. Many animal species seem to know a lot about the world of objects by instinct before they experience it. Insects play successful hide-and-seek with predator and prey before they have time to learn. Migrating birds use patterns of the stars to guide them over featureless oceans though they have never seen the sky before. How is this possible, if the empiricists are right that all knowledge is derived from experience?

Very fast learning is easily confused with immediately available innate knowledge. Many authorities on language development believe that much of the 'deep' structure of natural language must be inherited, because it seems impossible for a child to learn so much so soon. The crucial, not ethically accepted experiment was attempted by King James IV of Scotland (1473–1513), by marooning two infants on the island of Inchkeith, to be looked after by a dumb woman. It was reported (if not universally believed) that the babies spontaneously spoke Hebrew with a Scottish accent!

For 2000 years metaphysicians upheld their claim to innate knowledge by pointing to mathematics—especially to the spatial properties of Euclidean geometry—where new facts were discovered not by observations or experiments, but from studying diagrams and juggling with symbols. Only since the great German mathematician Karl Gauss (1777–1855) has it become clear that mathematical discoveries are of a special kind; constituting knowledge not of objects but rather of possible and impossible patterns. Gauss was among the first to realize that there is not only one possible geometry— Euclidean—but that other consistent geometries can be invented. So it is an empirical question to ask which axioms of suggested geometries best fit our world. (And it has been suggested that visual space is non-Euclidean.) The principal support for *a priori* knowledge of objects in space collapsed with the invention of alternative geometries. It is now believed that neither mathematics nor logic give new facts about the world, in the sense that facts are discovered by observation. In short, knowledge starts from experience, though mathematics and logic are extremely important for testing and drawing conclusions from mental models which might fit reality.

Animals respond appropriately to many objects and situations upon first encounter—but this does not make them metaphysicians. For they are heirs, by inheritance, to knowledge won through many generations of ancestral disasters, by natural selection, which benefits individuals with the most appropriate behaviour, who survive to hand on useful knowledge through their genes. Genetic coding can become modified by natural selection to give inherited knowledge to individuals, but what the parents learn individually in their life-time is lost.

Animals very low on the evolutionary scale rely almost entirely on unlearned reflex reactions; but their range of behaviour is restricted, and they respond in stereotyped ways. Some insects, especially bees, do show perceptual learning; for example the location of the hive and where nectar is plentiful—which could not be known innately. Patterns of petals leading to nectar became built into the inherited bees' brain, and those lacking flower-vision die for lack of honey. So here is a mixture of innate and learned knowledge.

It is hard to establish what knowledge human infants are born with and what has to be learned. The difficulties are that there are strict limits to experiments that can be tried on human babies, and that infants have extremely limited co-ordinated behaviour. Until recently, almost all we knew of learning how to see has came from young animals. Now, however, there are safe and effective techniques for learning from babies what they can see. We will consider these new techniques and findings a little later, after looking at some physiological effects of experience on animals—and at some other matters—including what it is like to recover when adult from infant blindness.

Physiological changes

Recent experiments have been aimed at whether physiological 'feature detectors' (Figure 4.7) are simply given innately, or whether early experience affects them. Kittens have been reared in environments of vertical stripes, then tested for vision of vertical and other orientations. It has been found, especially by Colin Blakemore, that kittens living in a world of only vertical stripes appear to be blind to horizontal lines—and they lack horizontal feature detectors. Similarly, kittens denied vertical stripes do not have well organized vertical feature detectors. This suggests that feature detectors are not completely laid down at birth; but are developed—or 'tuned'—by visual stimulation encountered by the individual. This is important for considering optimal environments for babies, especially as it has been found that some innately given neural mechanisms degenerate with lack of stimulation. This is clearly so for the ability to see depth stereoscopically. In childhood there are 'critical periods' for learning how to see, and without suitable experience at the right time such visual skills can be lost forever. Early visual environment of babies is highly important for adult vision—so nursery wallpaper should be considered!

Adaptation to disturbed images

Displaced images

To discover mechanisms of perceptual learning, we may look at experiments on animals and humans fitted with optical systems of various kinds to modify the retinal image, and see whether eye and brain compensate or adapt to the changed input. This was first tried at the end of the nineteenth century in famous experiments by the American psychologist G. M. Stratton, on himself. But first, let's look briefly at animal experiments of this kind.

Inverting goggles placed on a monkey had the effect of immobilizing her for several days: she simply refused to move. When finally she did move it was backwards—a point of some interest as these inverting goggles tend to reverse depth perception. Similar experiments have also been tried in chickens and hens. Right–left reversing prisms were attached to the eyes of hens by M. H. Pfister, who observed their ability to peck grain. The hens' behaviour was severely disturbed, and they showed no real improvement after three months wearing the prisms. The same lack of adaptation has also been found in amphibia, by R. W. Sperry. With vision rotated through 180°, it was found that they would move their tongue in the wrong direction for food, and would have starved to death had they been left to fend for themselves. Similar results were also obtained by A. Hess with chickens wearing wedge prisms which did not reverse the images, but shifted them by 7° to the right or to the left. He found that the chickens would always peck to the side of the grain, and that they never adapted to the shift of the image caused by the wedge prisms. Hess concluded:

Apparently the innate picture which the chick has of the location of objects in its visual world cannot be modified through learning if what is required is that the chick learns to perform a response which is antagonistic to its instinctive one.

It seems clear from the various experiments that animals show far less adaptation to a shift or reversal of the image than do human observers. Indeed, only monkeys and humans show any perceptual adaptation to these changes.

Now let's look at the classical work of G. M. Stratton on inversion of the retinal image for a human observer. He wore inverting goggles for days on end—and was the first man to have retinal images that were *not* upside down! He devised a variety of lens and mirror systems including special telescopes mounted on spectacle frames so they could be worn continuously. These generally inverted both vertically and horizontally. Stratton found that when a pair of inverting lenses was worn giving binocular vision the strain was too great as normal convergence was upset, and this did not adapt to the situation. He therefore wore a reversing telescope on just one eye, keeping the other covered. When not wearing the inverted lenses he would keep both eyes covered, or stay in a dark room. He slept in the dark.

At first, objects seemed illusory and unreal. Stratton wrote (1896–7):

... the memory images brought over from normal vision still continued to be the standard and criterion of reality. Things were thus seen in one way and thought of in a far different way. This held true also for my body. For the parts of my body were felt to be where they would have appeared had the instrument (the inverting lens) been removed; they were seen to be in another position. But the older tactual and visual location was still the real location.

Later, however, objects would look almost normal.

Stratton's first experiment lasted three days, during which time he wore the 'instrument' for about 21 hours. He concluded:

I might almost say that the main problem—that of the importance of the inversion of the retinal image for upright vision—had received from the experiment a full solution. For if the inversion of the retinal image were absolutely necessary for upright vision ... it is difficult to understand how the scene as a whole could even temporarily have appeared upright when the retinal image was not inverted.

Objects only occasionally looked normal, however, and so Stratton undertook a second experiment with his monocular inverting arrangement, this time wearing it for eight days. On the *third day* he wrote:

Walking through the narrow spaces between pieces of furniture required much less care than hitherto. I could watch my hands as they wrote, without hesitating or becoming embarrassed thereby.

On the *fourth day* he found it easier to select the correct hand, which had proved particularly difficult:

When I looked at my legs and arms, or even when I reinforced my effort of attention on the new visual representation, then what I saw seemed rather upright than inverted.

By the *fifth day*, Stratton could walk around the house with ease. When he was moving around actively, things seemed almost normal, but when he gave them careful examination they tended to be inverted. Parts of his own body seemed in the wrong place, particularly his shoulders, which of course he could not see. But by the evening of the *seventh day* he enjoyed for the first time the beauty of the scene on his evening walk.

On the *eighth day* he removed the inverting spectacles, finding that:

... the scene had a strange familiarity. The visual arrangement was immediately recognised as the old one of pre-experimental days; yet the reversal of everything from the order to which I had grown accustomed during the last week, gave the scene a surprising bewildering air which lasted for several hours. It was hardly the feeling, though, that things were upside down.

One has the impression when reading the accounts of Stratton, and the investigators who followed him, that there is always something queer about their visual world though they have the greatest difficulty saying just what is wrong with it. Perhaps, rather than their inverted world becoming entirely normal, they cease to notice how odd it is until their attention is drawn to some special feature, when it does look clearly wrong. Thus writing appears in the right place in the visual field, and at first sight looks like normal writing, except that when they attempt to read it, it is seen as inverted, or at least it appears odd.

Stratton went on to perform other experiments, which though less well known are just as interesting. He devised a mirror arrangement, mounted in a harness (Figure 8.1), which visually displaced his own body so it appeared horizontally in front of him at the height of his eyes. Stratton wore this mirror arrangement for three days (about 24 hours of vision), reporting:

The different sense-perceptions, whatever may be the ultimate course of their extension, are organised into one harmonious spatial system. The harmony is found to consist in having our experience meet our expectations . . . The

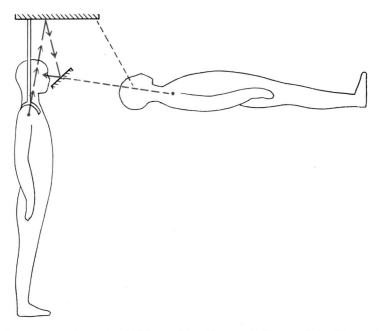

8.1 Stratton's experiment, in which he saw himself suspended in space before his eyes, in a mirror. He went for country walks wearing this arrangement.

essential conditions of the harmony are merely those which are necessary to build up a reliable cross-reference between the two lenses. This view, which was first based on the results with the inverting senses, is now given wider interpretation, since it seems evident from the later experiment that a given tactual position may have its correlated visual place not only in any direction, but also at any distance in the visual field.

Several investigators have followed up Stratton's work. G. C. Brown used prisms to rotate the field of both eyes through 75°, and found that this reduced the efficiency of depth perception; but there was little or no evidence of improvement with experience, though he and his subjects did find that they got used to their tilted world. Later, P. H. Ewert repeated Stratton's experiment using a pair of inverting lenses, in spite of the strain on the eyes found by Stratton. Ewert's work has the great merit that he made systematic and objective measures of his subjects' ability to locate objects. He concluded that Stratton somewhat exaggerated the amount of adaptation that occurred. This led to a controversy that is still unresolved.

The problem was taken up by J. Paterson and J. K. Paterson, using a binocular system similar to Ewert's. After 14 days they did not find complete adaptation to the situation. Upon re-testing the subject of the experiment eight months later, they found that when the subject wore the lenses again, he immediately showed the various modifications to his behaviour which he had previously developed while wearing the reversing spectacles. It seemed that the learning consisted of a series of specific adaptations, overlying the original perception, rather than a reorganization of the entire perceptual system.

The most extensive recent experiments on humans have been carried out by T. Erisman, followed by Ivo Kohler at Innsbruck. Both Stratton's and Kohler's experiments rely on verbal reports. Kohler stresses the 'inner world' of perception, following the European tradition which we find in the German Gestalt writers, and in the work of Michotte on the perception of causality (Chapter 4). This emphasis is foreign to the behaviourist tradition of America, and it is unfortunate that little precise recording of the subject's movements during the experiments was attempted. From the verbal reports it is difficult to imagine the 'adapted' world of the experimental subjects, for their perceptions seem to be curiously shuffled and even paradoxical. For example pedestrians were sometimes seen on the correct side of the street, when the images were right–left reversed, though their *clothes* were seen as the wrong way round! The suggestion is that having to avoid bumping into people produced re-learning of their positions on the pavement, but not of which side the buttons were on their coats. Writing is one of the more puzzling things observed. With right–left reversal a scene would come to look correct, except that at least sometimes writing remained right–left reversed and hard to read.

Touch had important effects on vision: during the early stages of adaptation objects would tend to look suddenly normal when touched, and they would also tend to look normal when the reversal was physically impossible or highly unlikely. For example, a candle would look upside down until lighted, when it would suddenly look normal—the flame going upwards, instead of downwards. Touch, even with a long stick, would switch the world the right way up.

There is later evidence, mainly from the work of Richard Held and his associates, particularly Alan Hein at Massachusetts, to show that for compensation to displaced images to occur, it is important for the subject to make active corrective movements. Held considers that active movement is vital for perceptual learning in the first place, as well as for compensation. An experiment with kittens is particularly ingenious and interesting. They brought up a pair of kittens in darkness; they could see only in the experimental situation, in which one kitten served as a control for the other. The two kittens were placed in baskets attached to opposite ends of a pivoted beam, which could swing round its centre, while the baskets could also rotate. It was arranged that a rotation of one basket caused the other to rotate similarly (Figure 8.2). With this ingenious device both kittens received much the same visual stimulation, but one was carried *passively* in its basket; the other, whose limbs could touch the floor, moved the apparatus around *actively*. Held and Hein found that only the active kitten gave evidence of perception, the passive animal remaining for a time effectively blind. But is this 'blindness' the absence of correlations built up between its vision and its behaviour? Or could the kitten indeed be seeing, but be unable to let us know that it sees?

Richard Held also undertook experiments on humans using deviating prisms, finding that active arm movement (striking a target with the finger) is necessary for effective adaptation. Is this adaptation *perceptual*, or is it *proprioceptive*—in the control system of the limbs? The principal supporter of proprioceptive adaptation is C. S. Harris. This cannot apply to adaptation to some kinds of distortion of vision, but in these cases the role of feedback from experience is less clear.

Distorted images

We have considered experiments on inverting and tilting the eye's images, but other kinds of disturbance can be produced, which are

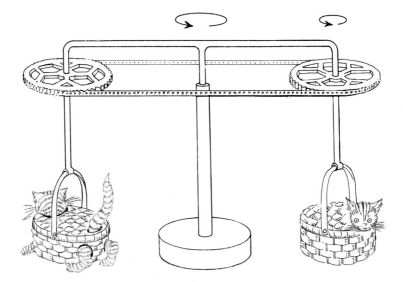

8.2 Apparatus designed by Held and Hein to discover whether perceptual learning takes place in a passive animal. The kitten on the right is carried about passively by the active kitten on the left. They thus have similar visual stimulation. Following visual experience limited to this situation, only the active animal is able to perform visual tasks—the passive animal remains effectively blind, until some time after the experiment.

important because they involve internal reorganization in the perceptual system itself, rather than relatively simple changes in the relation between the worlds of touch and vision. These can be investigated by wearing lenses which *distort*, rather than displace, the image on the retina.

J. J. Gibson found, while undertaking an experiment wearing prisms to deviate the field 15 degrees to the right that the *distortion* of the image, which such prisms produce in addition to the shift, gradually became less marked while he wore them. He went on to make accurate measures of the adaptation to the curvature produced by the prisms, and found that the effect diminished although his eyes moved about freely. Surprisingly, the adaptation was slightly more marked with free inspection of the figure with eye movements than when the eyes were held as still as possible. This adaptation does not depend on touch, so may be different from seeing *where* things are.

There is another kind of adaptation which at first sight seems similar to Gibson's with his distorting prisms, but is almost certainly different in its origin and its significance to the theory of perception. These effects are known as *figural after-effects*. They have received a great deal of experimental attention, though they remain somewhat mysterious. Figural after-effects are induced when a figure is looked at for some time (say half a minute) with the eyes held very still. If a curved line is fixated, a straight line viewed immediately afterwards will, for a few seconds, appear curved in the opposite direction. The effect is similar to Gibson's, but for figural after-effects it is essential that the eyes should be still. This seems to be a visual 'normalizing' process, useful for correcting standing errors such as astigmatism.

Using television to study visual disturbance

There are limitations to the kinds of inversion possible by simple optical means. A technique has been developed by K. U. Smith and W. M. Smith, using a television camera and monitor arranged so that the subject sees his own hand, displaced in space or time. It is a simple matter to give either right–left or up–down reversal of the image, and eye and hand movements are not affected. In this arrangement, the hand is placed to the side of the subject, behind a curtain so it cannot be seen directly. (Since the apparatus was far from portable these studies were limited to short experimental sessions.) In addition to reversals, the camera may be placed in any position, giving a view displaced in space. Using various lenses and camera distances, the size may be varied, and distortions may be introduced (Figure 8.3).

These techniques show that pure up–down reversal generally proves more disturbing than left–right reversal, and combined up–down and left–right is less disturbing than either alone. Changes in size or position had practically no effect on ability to draw objects, or on handwriting.

Displacement in time

An elaboration of this television technique makes it possible to displace retinal images, not only in space but in time. The original method was to introduce a videotape recorder with an endless tape loop, so that there is a time delay between the recording from the camera and play-back to the monitor. The subject thus sees his hands (or any other object) in the past, the delay being set by the gap between the record and play-back heads (Figure 8.4).

This situation is not only of theoretical interest but is also of practical importance, because controls used in flying aircraft and operating many kinds of machine have a delay in their action. If delay upsets the control skill this could be a serious matter. It was found

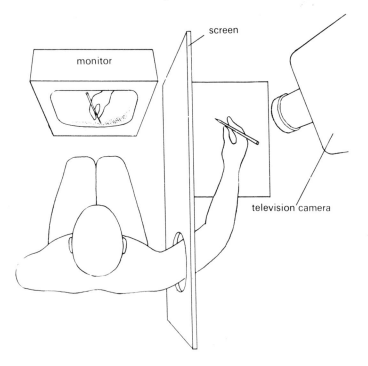

8.3 K. U. Smith and W. M. Smith's experimental arrangement using a television camera and monitor to vary the viewing position and size of the subject's own hands. He or she can draw or write very well despite effectively displaced eyes.

that a short delay (about 0.5 seconds) made movements jerky and ill co-ordinated, so that drawing became almost impossible and writing quite difficult (Figure 8.5). Practice gives little or no improvement. Present-day video equipment would make such experiments much easier to do.

Links of adaptation

A remarkable discovery was made by Ivo Kohler, wearing glasses which did not distort, but which were coloured half red and half green, so that everything looked red while looking to the left, and green when looking to the right (Figure 8.6). He found that the colours gradually weakened, and when the glasses were removed things seen with the eyes directed to the right looked red, and to the left green. This is quite different from normal after-images, for the effect is not

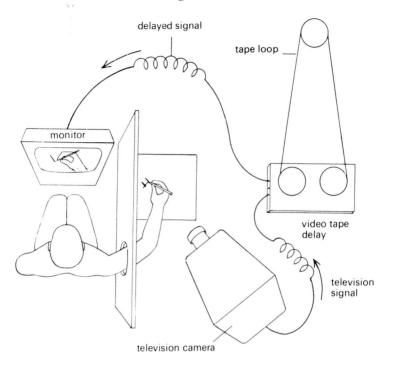

delayed signal

tape loop

monitor

video tape delay

television signal

television camera

8.4 Introducing time delay between action and seeing. The delay was given by the tape loop of the video recorder. This would now be done electronically.

related to the position of images on the retina, but to the position of the eyes in the head—so it must be due to compensation taking place in the brain.

This might be related to a striking effect discovered by Celeste McCollough in 1961. When lines of one orientation are repeatedly shown as red, and alternated with a green grating at another orientation, black and white lines of the same orientations appear coloured. The induced colours seen in the black and white lines are the complementaries of the adapting colours, for each orientation. This also works with movements, and to other visual stimuli, though not for non-visual stimuli such as sounds. So this is only within-vision learning.

Such *contingent after-effects*, as they have come to be called, show many of the characteristics of classical Pavlovian conditioning. They build up gradually with repeated stimuli; they decay rapidly when elicited by the 'unconditioned stimuli' (the differently oriented black and white lines). Left to themselves, the effects can persist for many

8.5 Drawing with a time delay. Left to right: normal; with TV but no delay; with TV and delay. The delay provides an insuperable handicap, though displacement in space can be compensated. (The result is of practical importance, since many controlled tasks, such as flying, do impose a delay between action and result.)

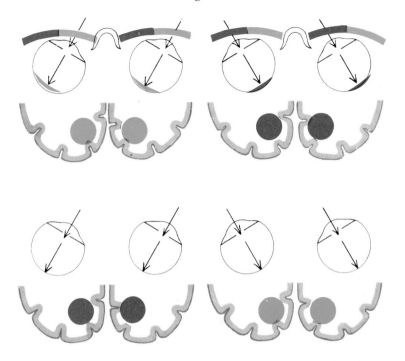

8.6 Demonstrating contingent after-effects. After directing the eyes through the green filter to one side, and then the red filter to the other, they become adapted and compensate to the filters for each viewing position. When the filters are removed, the side which was green looks red and vice versa. This adaptation must be in the brain, not the eyes.

hours, and even days or longer. It has also been suggested that links between related stimuli given by 'double duty' cells are teased out by this procedure. But it is now clear that highly 'artificial' and extremely unlikely pairings of stimuli can occur; so this is not a plausible explanation, for we should not expect cells to be pre-wired for highly unlikely combinations of stimuli. It is most likely that this is a kind of perceptual learning. Why, though, should it always be 'negative'? It seems to be recalibrating vision—to produce *constancy* of perception against irrelevant repeated sensory signals. This may be similar to compensating for new spectacles and such other visual learning. The McCollough effect is a dramatic and an experimentally useful phenomenon—illustrating how vision scales and adapts itself, generally avoiding its own errors, to make sense of the world. But generally useful processes can sometimes generate curious and revealing illusions.

Cultural differences

Do people brought up in different environments come to see differently? The Western world has visual environments with many straight parallel lines, such as roads, and right-angular corners of buildings and furniture and so on. These are strong, generally reliable perspective cues to distance. We may ask whether people living in other environments where there are few right angles and few long parallel lines have somewhat different perception. Fortunately several studies have been made of people living in such environments.

Those who stand out as living in a non-perspective world, were the Zulus. Their world has been described as a 'circular culture'—traditionally their huts were round, they did not plough their land in straight furrows, but in curves, and few of their possessions had corners or straight lines (Figure 8.7). They are thus ideal subjects for

8.7 Circular culture of the Zulus. They experience few straight lines or corners, and are not affected by 'perspective' illusion figures to the same extent as people brought up in a 'rectangular' Western culture.

our purpose. It is found that they experienced the Muller–Lyer arrow illusion (Figure 10.16) to only a small extent, and were hardly affected at all by other such distortion illusion figures.

Studies of people living in dense forest have also been made. Such people are interesting in that they do not see many distant objects, because they live in quite small clearances in the forest. When they are taken out of their forest, and shown distant objects for the first time, they see these not as far away but as small. (They have even reported that cattle look like insects.) People living in Western cultures experience similar distortion when looking down from a height. From a tall building, objects look much too small. Steeplejacks, and men who work on the scaffolding and girder structure of skyscrapers, are reported to see objects below them without this distortion. Again, active movement and handling of objects seem to be very important for calibrating the visual system.

Jan Deregowski has found that Zulus who do not experience illusions also see little or no perceived depth in these figures (using the depth measuring technique of Pandora's box (Figure 10.23)). So there seems to be rather clear evidence for cultural factors in these distortions, related to distance cues as available in the environment. (For a fuller discussion see Chapter 10.)

Recovery from blindness

It might seem a simple matter to bring up animals in darkness—to deny them vision for months or years—and then discover what they see given light. Pioneering experiments of this kind were undertaken by A. H. Reisen. He found severe behavioural losses; but some of these might be due to degeneration of the retina (which was found to occur in darkness, though less so with diffusing goggles) and also to the remarkably passive state of animals, especially monkeys, reared in the dark. It was difficult to infer specific perceptual changes, or losses, because of the general lack of behaviour of these animals. It is not socially possible to bring up human babies in the dark; but there are cases of adult recovery from blindness. Can these tell us how human perception develops?

Perception of the blind was described by Descartes in the *Dioptrics* (1637), as discovering the world by poking with a stick:

. . . without long practice this kind of sensation is rather confused and dim; but if you take men born blind, who have made use of such sensations all their life, you will find they feel things with such perfect exactness that one might almost say they see with their hands.

The implication is that this kind of learning might be necessary for the normal child to develop his or her world of sight.

The English philosopher John Locke (1632–1704) received a celebrated letter from his friend William Molyneux, posing the question:

Suppose a man born blind, and now adult, and taught by his touch to distinguish between a cube and a sphere of the same metal, and nighly of the same bigness, so as to tell, when he felt one and the other, which is the cube, which is the sphere. Suppose then the cube and the sphere placed on a table, and the blind man made to see; query, Whether by his sight, before he touched them, he could now distinguish and tell which is the globe, which the cube? To which the acute and judicious proposer answers: 'Not. For though he has obtained the experience of how a globe, and how a cube, affects his touch; yet he has not yet attained the experience, that what affects his touch so or so, must affect his sight so or so.'

Locke comments (in *Essay concerning human understanding*, 1690) as follows:

I agree with this thinking gentleman, whom I am proud to call my friend, in his answer to this problem; and am of the opinion that the blind man, at first, would not be able with certainty to say which was the globe, which the cube.

Here was a suggested psychological experiment—with a guessed result. George Berkeley (1685–1753), the Irish philosopher, also considered learning how to see in this kind of way:

In order to disentangle our minds from whatever prejudices we may entertain with the relation to the subject in hand nothing is more apposite than the taking into our thoughts the case of one born blind, and afterwards, when grown up, made to see. And though perhaps it may not be an easy task to divest ourselves entirely of the experience received from sight so as to be able to put our thoughts exactly in the posture of such a one's: we must, nevertheless, as far as possible, endeavour to frame conceptions of what might reasonably be supposed to pass in his mind.

Berkeley goes on to say, that we should expect such a man not to know that anything was:

high or low, erect or inverted . . . for the objects to which he had hitherto used to apply the terms up and down, high and low, were such only as affected or were some way perceived by his touch; but the proper objects of vision make new sets of ideas, perfectly distinct and different from the former and which can by no sort make themselves perceived by touch.

Berkeley gives his opinion that it would take some time to learn to associate touch with vision. But guessing, even by the most distinguished philosophers, is no substitute for observation and experiment. There have been several actual cases of the kind imagined by Molyneux and discussed by Locke and Berkeley. The most famous is that of a 13-year-old boy, described in 1728 by the celebrated eye and general surgeon, William Cheselden. (Cheselden was spokesman of the London surgeons. He attended Sir Isaac Newton in his final illness.) Although there are many reported cases, few have sufficient evidence of lack of vision in infancy, and early operations for removing congenital cataracts did not give a good retinal image for weeks or months, if then. The first reported case dates from AD 1020. There are a few recent cases where sight has been restored to an adult who was effectively blind from early infancy or birth, giving a good retinal image immediately.

Some of the reported cases are much as the empiricist philosophers expected. The patients could see but little at first, being unable to name or distinguish between even simple objects or shapes. Sometimes there was a long period of training before they came to have useful vision, and indeed in many cases it was never attained. Some gave up the attempt, reverting to a life of blindness, often after a period of severe emotional disturbance. It is important to note, however, that the reported cases do not all show extreme difficulty or slowness in coming to see. We should also remember that the early operations disturbed the optics of the eye, so there could not be a useful image until the eye had considerable time to settle down. This is particularly important in the case of removal of the lens for cataract of the lens, which constitutes all the earlier cases. The other kind of operable blindness—opacity of the cornea—involves less change or damage and far more rapidly gives an adequate retinal image. We shall now discuss in some detail such a case, which I had the good fortune to investigate, with my colleague Jean Wallace, starting in 1961. Following a life of blindness, S.B. was given successful corneal transplants when he was 52.

The case of S.B.

S.B. was an active, intelligent man, who spent a lot of time and energy imagining the sighted world. For many years he tried to get corneal grafts, but they were in short supply before corneal banks were set up, so it was not considered worth the risk. This is why he was blind from early infancy (or birth) until middle age, which makes this case most unusual.

While blind, S.B. would go for cycle rides, with a friend holding his shoulder to guide him; he would often dispense with the usual white stick, sometimes walking into obstructions such as parked cars, and quite often hurting himself. He liked making things with simple tools in a shed in his garden. All his life he tried to picture the world of sight: he would wash his brother-in-law's car, imagining its shape as vividly as he could. He longed for the day when he might see. Finally the attempt was made when corneal banks were set up, and it was successful. But though the operation was a success his story ended tragically.

When the bandages were first removed from his eyes, so that he was no longer blind, S.B. heard the voice of the surgeon. He turned to the voice, and saw nothing but a blur. He realized that this must be a face, because of the voice, but he could not see it. He did not suddenly see the world of objects as we do when we open our eyes. But within a few days he could use his eyes to good effect. He could walk along the hospital corridors without recourse to touch; he could even tell the time, from a large wall clock, having all his life carried a pocket watch, with no glass, so that he could feel the time by touching its hands, as he demonstrated to us with great skill. In the hospital he would get up at dawn, and watch from his window the cars and trucks pass by. He was delighted with his progress.

When S.B. left the hospital we took him to London and showed him many things he never knew from touch, but quite soon he became curiously dispirited. At the zoo, he was able to name most of the animals correctly, having stroked pet animals when a boy, and enquiring how other animals differed from the cats and dogs he knew by touch. (We learned of his early life from an older sister.) S.B. was also, of course, familiar with toys and models. He certainly used his previous knowledge from touch and reports from sighted people to help him name objects by sight, which he did largely by seeking their characteristic features. But he found the world drab, and was upset by flaking paint and blemishes. In particular, he looked at a lamp post, walked round it, studying it from different aspects, and wondered why it looked different yet the same. But at that time, he said that he noted more and more the imperfections in things, and would examine small irregularities and marks in paintwork or wood, which he found upsetting, evidently expecting a more perfect world. He liked bright colours, but became depressed when the light faded. His depression became marked and general. He gradually gave up active living, and three years later he died.

Depression in people recovering sight after many years of blindness is a common feature of these cases. Its cause is probably complex, but in part it seems to be a realization of what they have missed—not only visual experience, but opportunities to do things denied them during the years of blindness. Some of these people revert very soon to living without light, making no attempt to see. S.B. would often not trouble to turn on the light in the evening, but would sit in darkness.

We tried to discover what his visual world was like, by asking him questions and giving him various simple tests. While still in the hospital, before he became depressed, he was most careful with his judgements and his answers. We found that his perception of distance was peculiar, and this was also true of earlier cases. He thought he would just be able to touch the ground below his window, with his feet, if he hung from the sill with his hands; but in fact the distance down was at least ten times his height. On the other hand, he could judge distances and sizes quite accurately when he knew the objects by touch and, it seemed, especially distances from walking.

Although his perception was demonstrably peculiar, he seldom expressed surprise at anything he saw. He drew the elephant (Figure 8.8) just before we showed him one at the zoo; but upon seeing it, he said immediately: 'There's an elephanty', and said it looked much as he expected it would. On one occasion he did show real surprise, and this was an object he could not have known by touch—the moon. A few days after the operation, he saw what he took to be a reflection in a window. (He was for the rest of his life fascinated by reflections in mirrors, and would spend hours sitting before a mirror in his local public house, watching people.) But this time what he saw was not a

8.8 SB's drawing of an elephant. He drew this before having seen one. Half an hour later we showed him a real elephant, at the London Zoo. He was not at all surprised by it.

reflection, but the quarter moon. Thinking it was a reflection in the window, he asked the Matron what it was. When she told him— he said he had thought the quarter moon would look like a quarter piece of cake! He seldom, however, appeared surprised by such anomalies—including his dog-like drawing of an elephant.

S.B. never learned to read by sight (he read Braille, having been taught it at the blind school), but we found, in the hospital, that he could immediately recognize capital letters and numbers by sight. This so surprised us we could hardly believe he had been blind. It turned out (from the school records) that he had been taught upper case, though not lower case, letters and numbers by touch at the blind school. The children were given inscribed letters on wooden blocks, to learn by active touch. The children were not given lower case letters to learn, presumably because they were seldom used for inscribed brass plates, and so on. Although he read upper case letters immediately by sight, it took him a long time to learn lower case letters, and he never managed to read more than simple words. Now this finding that he could immediately read letters visually, which he had already learned by touch, clearly showed that he was able to use previous touch experience for his new-found vision. This is interesting to the psychologist, for it indicates that the brain is not so departmentalized as sometimes thought. There seems, indeed, to be a general knowledge-base available to all the senses. But this makes cases such as S.B.'s hard to apply to the normal case of a human infant coming to see. For the blind adult knows a great deal about the world of objects, through touch and hearsay, and can use his or her knowledge to identify objects from the slightest cues. It is also necessary to trust the unproved, not altogether reliable new sense, which means giving up the habits of many years and risking danger and humiliation. S.B. was teased by his friends for his mistakes, especially as he could not identify people at all reliably from their faces. So his situation was really very different from a child learning to see.

S.B.'s use of early touch experience comes out clearly in drawings which he did for us, starting while still in the hospital and continuing for a year or more. The series of drawings of buses (in Figure 8.9) illustrate how he was unable to draw anything he did not already know by touch. In the first drawing, the wheels have spokes, and spokes were a distinctive touch feature of wheels. The windows seem to be represented as he knew them by touch, from the inside. Most striking is the complete absence of the front of the bus, which he would not have been able to explore with his hands, and which he was still unable to draw six months or even a year later. The gradual introduction of writing in the drawings indicates visual learning: the sophisticated script of the last drawing meant nothing to him for nearly a year after the operation, although he could recognize capital letters while still in the hospital, having learned them previously from touch. It seems that S.B. made immediate use of his earlier touch experience, and that for a long time his vision was limited to what he already knew.

We saw, in a dramatic form, the difficulty that S.B. had in trusting and coming to use his vision when crossing a busy road. Before the operation he was undaunted by traffic. We were told that previously he would cross roads alone, holding his arm or his stick stubbornly before him, when the traffic would subside as the waters before Christ. But after the operation it took two of us, on either side, to force him across: he was terrified, as never before in his life.

When he was just out of the hospital, and his depression was but occasional, he would sometimes prefer to use touch alone when identifying objects. We showed him a simple lathe (a tool he had wished he could use) and he was very excited. We showed it him in a glass case at the Science Museum in London, then we opened the case so that he could touch it. With the case closed, he was quite unable to say anything about it, except that the nearest part might be a handle (which it was—the transverse feed handle), but when he was allowed to touch it, he closed his eyes and placed his hands on it, when he immediately said with assurance that it was a handle. He ran his hands eagerly over the rest of the lathe, with his eyes tight shut for a minute or so; then he stood back a little, and opening his eyes and staring at it, he said: 'Now that I've felt it I can see'.

More recently some patients have had their sight restored by acrylic lenses. Rejection by the eye's tissue was avoided by placing the lens in a tooth extracted from the patient, and implanting the tooth in the eye with the lens in a hole drilled through the tooth. There are now suitable materials for lenses which are not rejected.

Many of our findings with S.B. have been confirmed with these new cases. The extraordinary operations, from which people walk around with a tooth in their eyes as in a Greek myth, were performed in Italy, one of the patients being a philosopher. The results were reported in 1971 by the Italian psychologist Alberto Valvo. He, too, found that earlier experience of letters by touch gave immediate visual recognition. There are now cases of children in Japan who are being intensively studied, and Oliver Sacks has recently described a new case remarkably similar to S.B.

Although there are many philosophers and psychologists who think

(a)

8.9 (a) SB's first drawing of a bus (48 days after the corneal graft operation giving him sight). All the features he drew were probably known to him by touch. The front, which he could hardly have explored by touch, is missing, and he could not add it when we asked him to try. (b) Six months later. Now he adds writing. The inappropriate spokes of the wheels have been rejected, but he still cannot draw the front of the bus. (c) A year later, the front is still missing. The writing is sophisticated, though he could hardly read.

that these cases can tell us about normal perceptual development in children, I am inclined to think that they tell us rather little directly. As we have seen, the difficulty is essentially that adults, with their great store of knowledge from the other senses and reports from sighted people, are very different from the infant, who starts with no knowledge from experience. It is extremely difficult, if not impossible, to use these cases to answer Molyneux's question in terms of what babies see. The cases are interesting and dramatic, but when all is said they tell us little about the world of the infant, for adults with restored vision are not living fossils of babies. The philosophers did not guess that there would be transfer of knowledge, gained by touch, to new-found vision. So we have learned something surprising and interesting from these rare dramas.

Perception and behaviour

We know far too little about relations between perception and behaviour. For example, do *visual* distortions correspond with *behavioural* errors? Recent experiments show that they can be separate. Thus the visual illusory size change of circles in the Titchener illusion (Figure 8.10) does not correspond to how discs are grasped with the hands.

(b)

(c)

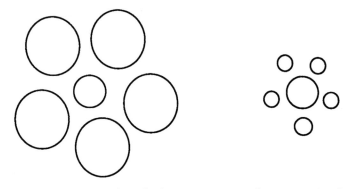

8.10 The Titchener illusion with touch. The central circles are the same size; but the circle enclosed by the large circles looks smaller. For the subject's hands, picking them up guided by vision, there is no distortion. Visual illusions do not always correspond with errors of behaviour.

Here vision is distorted though hand control is not. As David Milner and Melvyn Goodale said, in 1995, in *The visual brain in action* (pages 163–4):

The primary purpose of perception is to identify objects and places, to classify them, and to attach meaning and significance to them, thus enabling later responses to them to be selected appropriately. As a consequence, perception is concerned with the enduring characteristics of objects so that they can be recognised when they are encountered again in different contexts.

So vision of the object world is not necessarily tied to the observer's body. But to be useful the neural processes for behaviour must be geared very closely to the body, and to the here and the now. So perception and behaviour do not occupy quite the same worlds.

What do babies see?

Of the most immediate interest is what infants—babies in the first two years of life, without speech—see immediately and in a few months come to see. It is hard to get reliable data because at first there is little motor co-ordination and no language. Recent experiments have, however, shown that babies at birth do have considerable vision. Within hours they can copy facial expressions of the mother, especially sticking out the tongue, so they must start out with some ability to see. Through, and following, the pioneering work of the Swiss developmental psychologist Jean Piaget (1896–1980), there are now several safe and effective techniques for revealing what babies see and under-stand through the first months and early years of human life. Piaget invented this experimental epistemology.

The first question is: do babies see objects as permanent things, in a world shared by other children and adults?

The Bishop Berkeley effect—permanence of objects

We have stressed throughout that perception is far more than simply responding to stimuli. Rather, it is acting appropriately to assumed sources, or causes, of stimuli. So we must consider not only *pattern* recognition but also *object* recognition—objects having lawful pasts and futures, and unsensed characteristics which often are vital for behaviour.

A primary characteristic of objects is their permanence, though they appear but fleetingly from time to time, as we glance at them, or as they reappear from being hidden by nearer objects. How do infants cope with objects disappearing? We may ask first: how do philoso-phers cope with disappearing objects? Bishop George Berkeley (1685–1753) suggested that objects only exist while they are perceived. This is unforgettably expressed in Ronald Knox's famous limerick and its reply:

> There was a young man who said, 'God
> Must think it exceedingly odd
>   If He finds that this tree
>   continues to be
> When there's no one about in the Quad.'

God replies:

> Dear Sir,
>   Your astonishment's odd:
> I am always about in the Quad.
>   And that's why the tree
>   Continues to be.
> Since observed by
>   Yours faithfully GOD

The English empiricist philosopher John Stuart Mill (1806–73) described objects as 'permanent possibilities of sensation'. Is this how infants see them?

Piaget suggested that very young babies do not have any notion that objects continue to be when not observed. Then they discover that when part of an object is revealed the rest will be there, to be found, and finally that whole objects continue to exist when they are not seen. Piaget held that these are discoveries made by active exploration, though this has been questioned because baby behaviour is so clumsy that such experiments must be hard to carry out in the first few months of life. (The alternatives—that the information is out there to be 'picked up', or that there are innate principles for completing missing objects—seem, however, more implausible.)

It is found that young babies recognize objects in spite of changes of view. They show 'object constancy'; so they do not respond to appearances, but rather to objects as existing and remaining the same in spite of considerable change of appearance. Only slowly do they come to see themselves as moving in a room. Recognition of self is not developed before about two years of age. Part of the evidence for this, is behaviour to their own image in a mirror. When a baby touches his or her own face—not the mirror—it seems that the baby has some notion of self.

Techniques for studying infant vision

How are stages of learning how to see discovered? Techniques include looking for prediction, surprise, boredom.

Prediction

This can mean simply expecting what has happened to continue happening, as for Pavlovian conditioning or it may indicate cognitive understanding, including predicting others' behaviour in novel situations. It is said that children develop 'theories of other minds', tested by predictions which may or may not be confirmed. Perhaps the strongest evidence for other minds is fibs; such as telling a child there are no cookies in the tin, but he or she finds that there are! The fib, or lie, is very different from the behaviour of inanimate objects; though babies only slowly come to distinguish between living and non-living.

Surprise

A key sign of prediction, and understanding is surprise. Cognitive perception is inherently predictive—depending on knowledge and assumptions—revealed most dramatically by failed predictions, with accompanying surprise. There are various signs of surprise in children before speech, such signs as a sudden agitation, and increased heart rate. Almost from the start, babies have expectations of objects. Expectations, and so surprises, become richer and more varied as perception develops. Surprise is, indeed, the principal reward of science.

The Scottish experimental psychologist T. G. R. Bower, projected images of objects stereoscopically (as in Figure 10.24) to appear a foot or so away, though there was nothing to touch. (It should be noted, however, that effective stereoscopic vision is not present before about three months.) Bower found that young infants show surprise when they reach out to grasp images having no substance. No doubt this is like our surprise at successful conjuring. When an interesting object such as a teddy bear is passed behind an opaque screen, and so disappears, the baby will move her eyes to the further side of the screen, evidently expecting it to re-appear. If it does not appear she is likely to get upset. What happens if the teddy gets hidden behind the screen, then another object, such as a toy fire engine, appears instead? Very young babies will show no surprise at a teddy turning into a fire engine; but at a year or so, there is surprise at such a transformation. So at about this age the baby possesses the knowledge that objects such as teddies do not turn into other objects such as fire engines.

In general, surprise at the unusual reveals knowledge of the usual. This is useful for seeing what babies see—and for reading minds.

Boredom

Signs of almost the opposite—boredom—are also useful. Babies get bored (become habituated) with repeated stimuli. But suppose the stimuli change but what they *mean* remains the same—would such repetition also produce boredom and habituation, so that interest flags and behaviour ceases? The trick is to discover whether the meaning of one pattern or shape is equivalent to another by habituating the baby (boring it) with the first, and noting whether it remains bored or habituated to the second.

There are other techniques for looking into baby vision and understanding.

Suck it and see

The American psychologist Jerome Bruner made imaginative use of one of the infant's few well organized behaviours, by getting infants to focus a slide projector by sucking on a special teat. In this way they would bring pictures of interest into focus.

What do babies prefer?

Even very young babies have quite well co-ordinated eye movements. These indicate what the baby is looking at, and so what is seen and preferred. Robert Fantz, the pioneer in using infants' eye movements to see what babies see, presented pairs of pictures and noted which was preferred from the times the baby looked at each picture. This technique has proved to be very useful, though a snag is the baby may find neither picture of interest; so much of what can be recognized or discriminated may be missed by the experimenter.

If one picture screen is blank, and the other has stripes (often called gratings), the infant will look more at the grating. There is a general preference for complexity. By making the grating more and more closely spaced until there is no preference, the baby's visual acuity can be measured. Infant acuity is considerably less than adult's, probably because the retina is not fully developed. It used to be thought that infants lack accommodation to different distances, but this is not too serious an experimental problem. Oliver Bradick and his wife Janette Atkinson carry out extensive clinical tests, using a technique based on the 'red eyes' seen in flash portraits when the sitter's eyes are not accommodated to the distance of the camera (cf. pages 42–44).

Very young babies can discriminate between orientations of coarse gratings, and between various simple outline shapes—cross, circle, triangle, square, acute and obtuse angles—with a general preference for patterns with curves.

Presented with a simple face-like picture, and the alternative of a jumbled face, Robert Fantz discovered that babies spend about twice as long looking at the normal face (Figure 8.11). This may imply innate recognition of faces; but early learning is not ruled out, as the mother's face had not been hidden from the baby. Figure 8.12 shows what happened.

Babies tend to prefer solid objects to flat representations of the same objects; so possibly they have some innate appreciation of depth. This is suggested by reaching and grasping of objects, some authorities believing that very young babies open their fingers to the size of an object before touching it. The famous visual cliff shows that babies at about four months are aware of potentially dangerous drops.

The visual cliff

Eleanor Gibson, while picnicking on the rim of the Grand Canyon, wondered whether a young baby would fall off the edge. This thought led her to a most elegant experiment, for which she devised a miniature and

---

8.11 (a) Fantz's apparatus for observing babies' eye movements while they are shown various designs, or objects. Here the baby is shown an illuminated ball, while the eye positions are photographed. (b) A simple face and a randomized face-like design, which were shown to very young babies. They spent longer looking at the typical face picture (as judged by their eye movements).

(a)

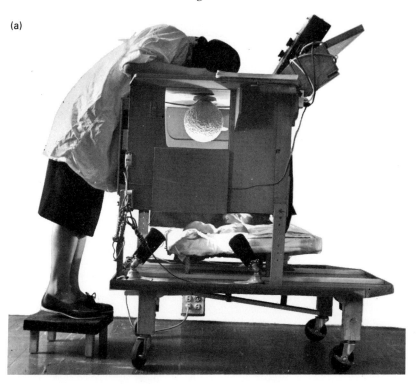

(b)

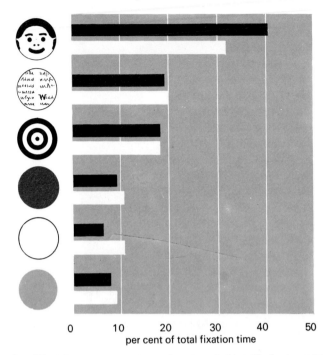

8.12 Results of Fantz's eye movement experiments on babies. The horizontal bars show the relative times they spent looking at the designs shown on the left of the diagram.

safe Grand Canyon. The apparatus is shown in Figure 8.13, which shows a central 'bridge' with on one side a normal solid floor, and on the other side a drop, covered by a strong sheet of glass.

An infant (or in other experiments a young animal) is placed on the central bridge. The question is: will the baby crawl over the drop? The typical answer is that the baby will not leave the bridge for the drop, and cannot be enticed over it by his mother shaking his rattle, though he will crawl quite happily on the normal floor. So it seems that babies at the crawling stage can appreciate a drop, probably from motion parallax when the baby moves its head, with innate knowledge of the danger of falling.

As adults, we cannot remember what it was like to enter the world and learn how to cope. Very soon we learn to control things; either by handling them, or by persuading others to do our bidding. So we may suppose technology derives from picking up dropped mugs, and using powers of symbols for persuading other minds (extended to controlling nature by magic) becomes how, as adults, we see and do from our early experiences as infants.

8.13 The 'visual cliff'. In this famous experiment designed by Eleanor Gibson and R. D. Walk, babies or young animals are tempted to a visual drop, protected by a strong glass sheet. The baby refuses to crawl on the glass over the drop, so evidently sees the depth, innately associated with danger.

Forgetting how to see

This chapter is called 'Learning how to see'. We have considered
several aspects, ending with the wonderful potentialities of human
infants—as babies learn to see and understand and, far more than any
other species, come to control the world. Sadly, some people forget
how to see. This is so for what are called visual agnosias.

The word 'agnosia' was coined by Freud, meaning lack of
knowledge for perception. It occurs with damage to regions of the
brain involved in relating sensory signals, which may be entirely
normal, to object knowledge. Agnosia may be only for vision, or for
touch or hearing. Agnosias have great psychological interest, for they
apply to inability to recognize objects though the eyes, or the ears or
whatever, are working normally. It is *meanings* of sensory signals that
are lost.

There can be specifically visual, or auditory, or touch agnosia—
when objects cannot be recognized by one of the senses though they
can by the others. Agnosias are associated with more or less specific
brain damage, (behind the central sulcus). The key is loss of connecting knowledge for object recognition; so it is hardly misleading to call
visual agnosias 'forgetting how to see'.

The onset can be sudden. The classical case was described by
Lissauer in 1890: an 80-year-old man (G.L.), who in a storm was blown
against a wooden fence, knocking his head. He found that everything
looked unfamiliar, and he mis-identified familiar objects—such as
seeing pictures on the wall as boxes, and confusing his jacket with his
trousers. He could, however, copy drawings; so his vision was
normal, except for loss of meaning.

The most dramatic and insightful accounts are those of the neurolo-
gist Oliver Sacks, especially in *The man who mistook his wife for a hat*.
His patient Dr P. was a gifted music teacher, who though retaining his
musical gifts, gradually forgot how to see. After Dr P. confused his
own foot with his shoe, Sacks showed him pictures in a magazine:

His responses were very curious. His eyes would dart from one thing to
another, picking up tiny features, individual features, as they had done with my
face. A striking brightness, a colour, a shape would arrest his attention and elicit
comment—but in no case did he get the scene as a whole. He failed to see the
whole, seeing only details, which he spotted like blips on a radar screen. He
never entered into relations with the pictures as a whole—never faced, so to
speak, its physiognomy. He had no sense whatever of a landscape or scene.

He would also 'see' things not in the picture: a river, a terrace, and
coloured parasols which were not there. Yet he could see and recog-

nize abstract shapes without difficulty. Faces were confused and expressions meaningless, though voices immediately identified the speakers. Then:

... he seemed to think he had done rather well. There was a hint of a smile on his face. He also appeared to have decided that the examination was over, and started to look round for his hat. He reached out his hand, and took hold of his wife's head, tried to lift it off, to put it on. He had apparently mistaken his wife for his hat! His wife looked as if she was used to such things.

The various agnosias of vision, touch, hearing, and so on, confirm that the senses are separately processed by specialized regions of the brain. Yet normally the senses come together, to give unified perceptions of objects. How this happens is far from understood. The S.B. case of recovery of blindness (pages 153–159) seems to show that there is transfer of knowledge from one sense to another; in this case it was from touch to vision, as following the eye operation for blindness he could immediately identify letters, and tell the time from earlier touch experience of letters and the hands of his watch. He had learned how to see before he had sight. People with visual agnosia have sight but have forgotten how to see.

9

Realities of art

The key to perception—to art and science, indeed to civilization—is *representing*. How concerned are artists with representing objects in two dimensions as though in the space of the three-dimensional world?

From the beginning, art seems to have been associated with magic and with religious, not only physical, views of the world. Cave paintings seem to have had significance in terms of what we would call sympathetic magic. Perhaps pictures were regarded as talismans; as charms for good or evil, capable of evoking thoughts and moods not only in men but also in Gods: for they are found on cave walls almost inaccessible to human view. Sometimes they were painted in succession, one upon another, as though the place of the paintings took on a special quality. Something of this magic quality of art is associated with all religious paintings, including the Christian, and is with us still.

In our society pictures are valued in monetary terms far beyond 'reason'. An original painting may cost hundreds even thousands of times more than a virtually indistinguishable copy. A painting's value depends greatly on the artist and whether it is identified as genuine. It is something of a sacred object even when its subject is profane. Evidently paintings are regarded as more than surrogate retinal images: they can be valued more highly than sight itself.

In religious paintings, the subject was of first importance, for they were used for conveying and unifying belief among largely illiterate people. But what is portrayed is not always important. A bunch of grapes or a dead hare are not in themselves interesting or valuable, and yet a picture of such worthless objects can command more than a man earns in a lifetime. But is this conceivable if paintings are no more than representations of the shapes and colours of objects? It is, indeed, just because a painting is not simply a spatial record that it can have a value and an interest far greater than the objects it portrays.

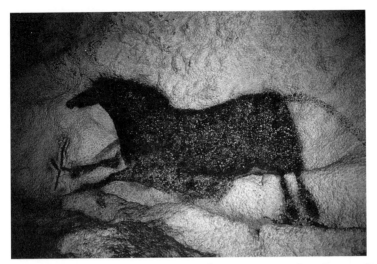

9.1 Galloping horse. This example of palaeolithic cave painting, from Lascaux (Dordogne, France), dates to approximately 15 000 BC. Photo: AKG London.

With the recent discovery of the Chauvet caves in the South of France, examples of pictures go back 30 000 years. Upon seeing cave art for the first time, Picasso is supposed to have said: 'Have we made any progress?' It would seem that over the millennia art, religion, and science have gradually separated. One might say that art is now primarily evocative—science primarily explanatory. They have grown ever further apart since the Greeks; but each depends in various ways upon the other, finding and giving mutual inspiration. An outstanding mutual contribution is the discovery or invention of rules for showing depth by perspective.

Representations of perspective

It is striking how seldom accurate representation of space is found in art. Indeed some of the most prized paintings appear perfectly flat. Certainly the ability to represent depth is regarded as but one, and a minor, accomplishment, except in special cases. And yet paintings generally include objects which we normally see in depth. So there is a problem: why is strict geometrical perspective rather seldom adopted? After all, the retinal image is in geometrical perspective (every feature halved in size with doubling of distance), so why does the artist seldom reproduce the eye's image to record depth in pictures? We may say at once that there is a very real difficulty in this. Although the retinal image is in strict perspective, like a photograph, this is not how we see the world—because the image is effectively modified by perceptual scaling (see Chapter 10).

Perspective is a remarkably recent development in Western art. In prehistoric pictures, and in pictures of early and non-Western societies, there was no formally correct perspective. It does not appear until the Italian Renaissance of the sixteenth century. We will consider this after looking at, to us, curious perspectives of Egyptian and oriental pictures.

Egyptian art

In the highly developed formalized painting of the ancient Egyptians, going back 5000 years, human figures are shown as anatomical impossibilities. Hands and feet and breasts are all shown as from the side, the shoulders front view and heads in profile, with an eye at the side, like an ear (Figure 9.2). In Egyptian pictures there is virtually no foreshortening or perspective. The sizes of the figures are determined not by distances but by their relative importance. The Pharaohs and the heads of families are huge compared with the lesser figures, such as conquered people and wives. Instead of an overall picture from a particular viewing point—which is familiar to us—each feature is represented from its most easily recognized view, combined to look paradoxical.

Almost all the stone statues stand or sit with no activity, looking straight ahead, with the head at right angles to the shoulders. This gives them a timeless dignity; but did the sculptors accept these restraints for art's sake, or because they lacked the artifice to show free natural positions and movements? Was it because massive stone simply did not allow such freedom to the sculptors? Or—very different—was it because the statues were based on drawings? It is known from part-finished colossal statues that they started as drawings on a grid of squares (the size of the statue's fist being marked on the block to set the scale) continually redrawn as stone was removed. (Wouldn't it have been possible to project a shadow of a wire grid onto the stone, so it would remain as the surface was cut away?) The many small figures of servants, made of wood or clay, placed in tombs to serve in the after life, are very different: far more naturalistic, free of the restraints of the formal reliefs and impressive stone statues. Is this because for making these small informal figures, drawings were not used?

9.2 Egyptian figures. The figure has each feature—eyes, shoulders, feet and so on—in characteristic 'sideways' positions, without perspective. (Courtesy of Andromeda Oxford Limited.)

Oriental art

Chinese drawings and paintings represent space with rules quite different from the geometry of the retinal image, often from many viewing positions at the same time. Chinese and Japanese paintings often combine a downwards view of buildings with a sideways view of people. Preferred for monumental pictures, was the large hanging scroll, two metres high and sometimes much more, especially from the tenth century. With a series of viewpoints, mountains could be shown at the top without dominating the lower parts of the pictures (Figure 9.3). The aim was to unify nature with man.

Here rectangular objects, such as tables, look odd to us as they have 'reversed' perspective—diverging rather than converging with distance. One has to be careful, though, for by a particularly interesting distortion illusion, parallel lines appear to converge when depth is suggested (Figure 10.15). Most of the 'negative perspective' in oriental painting is produced in this way—in the eye of the beholder—though sometimes it is exaggerated by the artist.

One might say that the ancient Egyptian combinations of views of parts of figures—profile heads, full-face eyes, sideways breasts, and so

9.3 Chinese perspective. This is odd to our eyes, as it is neither geometrical nor as the world appears through constancy scaling of the retinal image. The Chinese adopted elaborate symbolic conventions. What appear as drawn distortions ('negative perspective') are often in the observer's eye and brain as in Figure 10.21.

on—are much the same as the mixed views of Japanese and Chinese painting, though the 'units' are different. For the Egyptians the units are parts of objects; in oriental pictures the figures have a consistent viewpoint, but they are placed in a scene as views from displaced eyes.

Western Renaissance

It is an extraordinary fact that simple geometrical perspective took so long to develop—far longer than fire or the wheel—yet in a way it has always been present for the seeing, as the images in our eyes are perspective projections. Linear geometrical perspective in pictures is the invention of Italian Renaissance artists. It has had profound effects on art and on science.

How to paint a scene in the perspective of the eye's image was discovered by the Italian artist–engineer Filippo Brunelleschi (1377–1446). While planning the cathedral of Florence, he introduced a painting of his imagination into the existing buildings, to see how it would look when built. This virtual reality was accomplished with a painting with a hole in it and a mirror (Figure 9.4).

The laws and principles of perspective were first clearly described

A silver area on the panel reflected the sky

9.4 The first virtual reality. Brunelleschi introduced a painting of the proposed cathedral into the surrounding buildings, with a mirror. The painting of the proposed cathedral was about 0.5 × 0.5 metres, the mirror being half this size. Looking through a peephole in the painting at a mirror behind which were the existing buildings; the painting appeared as a 'virtual reality' of the future. This is worth trying, with a postcard and a mirror.

by Leonardo da Vinci (1452–1519) in his *Notebooks*, where he outlines a course of study for the artist. The problem was not only to present existing objects in perspective but also to create perspective for imaginary objects, when there was nothing to copy. Leonardo called perspective 'the bridle and rudder of painting', describing it in the following way:

Perspective is nothing else than the seeing of a plane behind a sheet of glass, smooth and quite transparent, on the surface of which all the things approach the point of the eye in pyramids, and these pyramids are intersected on the glass plane.

Leonardo treated the perspective of drawings as a branch of geometry. He described how perspective could be drawn directly on a sheet of glass (Figure 9.5), a technique used by the Dutch masters and, in a later form, with the camera obscura which employs a lens to form an image of the scene which may be traced directly (page 34).

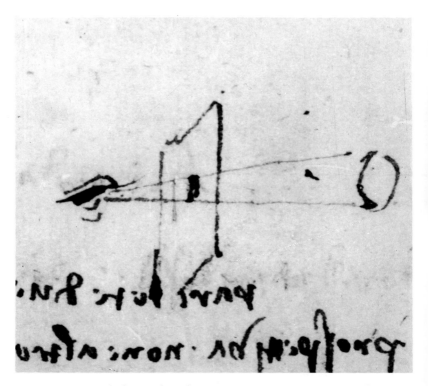

9.5 Leonardo's sketch showing how three-dimensional objects appear on a plane intersecting the cone of rays from object to eye. It is worth tracing a scene on a window pane.

The projection is determined simply by the geometry of the situation and this constitutes so-called geometrical perspective; but as Leonardo realized more clearly than many later writers, there is more to the matter than pure geometry. Leonardo includes in his account of perspective such effects as increasing haze and blueness with increasing distance, and the importance of shadows and shading in drawings to represent the orientation of objects. These considerations go beyond geometry. Geometrical (or linear) perspective was dominating Western painting by AD 1500 (Figure 9.6), the first artist to use it outside Italy being Albrecht Dürer (1471–1528), from Nuremberg. Spreading to Eastern art it could be incongruous. Later, tracings from the camera obscura were used by the Dutch painter Jan Vermeer (1632–75) for some of his interiors, by the Venetian Antonio Canaletto (1697–1768) (Figure 9.7), and many artists prior to photography, which became available after the middle of the nineteenth century, and is undoubtedly much used by draughtsmen and painters in the present time.

Let's look at what is going on, in as simple a way as possible. Consider a simple ellipse such as in Figure 9.8. This might represent an elliptical object seen normally or a circular object seen obliquely. This figure does not uniquely indicate any one kind of object; it could be a projection of any of an infinite variety of objects, each seen from a certain angle of view. The art of the draughtsman and painter is in large part to make us accept just one out of the infinite set of possible interpretations of a figure: to make us see a certain shape from a certain point of view. This is where geometry goes out and perception comes in. To limit the ambiguity of perspective, artists must make use of perceptual cues available to a single eye. They are forbidden the binocular distance cues of convergence and disparity, and also motion parallax. Indeed, these cues will work against them. Paintings are generally more compelling for depth when viewed with a single eye with the head kept still—because motion and the second eye show that the picture is flat.

We have to consider a double reality. The painting is itself a physical object, and our eyes may see it as such, flat on the wall; but it can also evoke quite other objects—people, ships, buildings—lying in a different space and, indeed, in a different time. It is the artist's task to make us reject the first reality while conveying the second—so that we see his world, not mere patches of colour and textures on a surface.

As we have seen from the example of the ellipse, a picture can represent a given object from one viewing position, or any of an infinite set of different objects seen in other orientations, or distances.

9.6 An early, very elaborate example of perspective: 'The Annunciation' by Criveli (*c.* 1430–95).

9.7 This Canaletto is a fine example of perspective. Has he painted the perspective as given in the image of the eye (or in a camera obscura, or photographic camera)—or following size–distance modifications by the size constancy of his visual brain? (And of course he has artistic licence to change it as he wishes.)

9.8 Is this shape seen as a perspective-depth drawing? Only if we know what it is.

This means that for the picture to represent something unambiguously we must see what the object really is—what its shape is, and how it lies in space. It is very much easier to represent familiar than unfamiliar objects. When we know what the object is, then we know how it must be lying to give the projection given by the artist. For example, if we know that the ellipse is representing a circular object, then we know that this must be lying at a certain oblique angle, the angle giving the eccentricity drawn on the plane by the artist. We all know that wheels, dinner plates, the pupil of the human eye, and so on, are circular objects, which makes the artist's task easier. This is clear from the power that very simple line drawings have for indicating form and orientation and distance—when the object is familiar. Consider the drawing of the boy with the hoop in the cartoon (Figure 9.9). It is clear that the ellipse represents a circle at an oblique angle, because we know that it represents a hoop and we know that hoops are circular. The hoop in this figure is, in fact, the same as the ellipse seen ambiguously, without context, in Figure 9.8. But when we know what it is, we know how to see it. It would have been extremely hard for the artist to represent a squashed hoop.

Look at the amoebic shape of the spilled wine in Figure 9.10: it is seen as lying on a flat surface (the road), though the shape alone could equally well represent an infinity of shapes, lying in various orientations. Suppose we remove the rest of the drawing so that we have no clue as to what it represents (Figure 9.11, showing only the puddle). It could equally well be anything of rather indeterminate shape standing up and facing us, perhaps something like an anvil. (Does it not look slightly higher in the full drawing where it is clearly lying obliquely

9.9 In this cartoon, the same elliptical shape is clearly a circular object, tilted.

on the ground, than when it appears alone as an upright shape?) Although the figure is so simple, a wiggly line, it is evoking vast experience of seeing and handling objects (including mis-handling when we drop bottles), and this physical knowledge very much determines how we see the amoeboid shape as a particular object.

We may now take another example, again of an ellipse in a cartoon, but this one illustrates a rather different point (Figure 9.12). This is of

9.10 The puddle is clearly lying flat on the ground—this is what puddles do!

9.11 The same shape as the puddle in the cartoon; but how does it lie in space now, without a context? It could be upright.

9.12 Another ellipse. This time we assume it is a circle and see it as flat, because we know that the invisible boy under the floor (almost seen!) would generally saw a roughly circular hole.

some interest, for it is presented without any explicit context and yet it clearly lies on the floor. It is seen as a circle. The boy below (not shown—yet seen!) could be cutting an elliptical hole, but we assume he is cutting a circle. This puts our viewing position at a certain height which is not determined by any other feature in the drawing, but only by our interpretation of the meaning of the shape, based on our knowledge of small boys.

Size constancy scaling

When an artist employs strict geometrical perspective he does not draw what he sees—he represents his retinal image. As we know, these are very different; for what is seen is affected by constancy scaling (see pages 171–172 for more details). A photograph, on the other hand, represents the retinal image but not how the scene appears. By comparing a drawing with a photograph taken from exactly the same position, we could determine just how far the artist adopts perspective and how far he or she draws the world as seen after the retinal images are scaled for constancy. In a photograph distant objects generally look too small. It is a common disappointment that a grand mountain range comes out as a row of pitiful molehills.

The situation here is curious. The camera gives true geometrical perspective; but because we do not see the world as it is projected on the retina, or in a camera, the photograph looks wrong. Indeed it is fortunate that perspective was invented before the photographic camera, or we might have had difficulty accepting photographs as other than weird distortions. As it is, photographs can look quite wrong, particularly when the camera is very close, or not held horizontally. Aiming a camera upwards to take in a tall building gives the impression of the building leaning backwards. Yet this is the true perspective.

Architects recognize that the visual compensation for distance is less efficient when looking upwards, and so have built columns and towers to diverge from bottom to top. The most notable example is the magnificent Campanile at Florence designed by Giotto. Here the artist has applied reversed perspective to reality, to compensate for the inadequate constancy scaling of viewers in this situation. There are examples of this on the horizontal plane, notably the Piazza San Marco in Venice, which is not a true rectangle, but diverges towards the cathedral so that it appears to be rectangular (and even larger) when the cathedral is viewed from across the Piazza. We find similar 'distortions' of reality to suit eye and brain in Greek temples; though the Greeks never discovered the projective geometry of perspective. (It is likely that they achieved good compromise results by trials with earlier wooden buildings. In any case, this is a quite different problem from perspective projections of pictures on a plain surface.)

In an important sense, perspective representations of three dimensions are wrong as they do not depict the world as it is seen; but rather, the (idealized) images on the retina. We do not see our retinal images—and we do not see the world according to the sizes or shapes of the retinal images—for these are effectively modified by constancy scaling. Should not the artist ignore perspective and simply draw the world as he sees it?

If the artist ignores perspective altogether, his painting or drawing will look flat, unless other cues to distance are utilized with sufficient force. This seems to be almost impossible; for if the artist did succeed in suggesting depth by other means, this picture would look wrong, for these cues would trigger the constancy scaling system, to expand more distant represented objects. Evidently this means that the artist should use perspective (draw distant objects correspondingly smaller) if the viewer's constancy scaling is affected by the depth cues. Indeed if *all* the normal depth cues were present *complete* perspective should be used, for the viewer would see sizes and distances as though seeing the original scene. But, and this is the important point, in fact, the artist can hardly hope to provide all the depth cues available from reality, so should use a reduced perspective for maximum realism. Something of this is illuminated by the experiments of Adelbert Ames:

The Ames demonstrations

The American psychologist Adelbert Ames II started life as a painter. He went on to create a series of brilliant demonstrations, the most famous being the distorted room and the trapezoid window.

The farther wall of the Ames room is tilted, so that it does not lie normal to the observer. 'Negative' perspective makes this oddly shaped room give the same retinal image as a normal rectangular room. There is an infinite set of 'distortions' which give the same images as those of a normal room; so there can be an infinite variety of Ames rooms looking like normal rooms, provided their design follows certain quite simple rules. The better it is designed and constructed, the more it looks like a normal rectangular room—ideally being indistinguishable. There is really nothing surprising in this, because the image it gives is the same as for an ordinary room, so it must look the same. But if objects are placed in the room odd things happen, which could not be predicted with certainty.

An object (such as a person) placed at the further back corner looks the same distance and smaller than the same-sized object (or person) at the nearer corner. It looks too small simply because the image is smaller; and the same distance, because there is no visual information that it is at a distance different from the other object. So an adult may be shrunk to appear smaller than a child (Figure 9.13).

Evidently we are so used to rectangular rooms, we accept it as axiomatic that it is the *objects* inside (the people) which are odd sizes, rather than the *room* being an odd shape. This is a betting situation; for either could be peculiar. Here the brain makes the wrong bet, as the experimenter has rigged the odds.

An interesting feature of the Ames room is its implication that perception is a matter of making the best bet on the evidence. It has been reported that young wives may not see their husbands as

9.13 The Ames room—impossible? We accept that the room is rectangular—though in fact it is not—and we see the figures as different sizes though they are the same. We are so used to rooms being rectangular that we bet on this room being normally shaped. Here we are wrong.

distorted in the room, but see them as normal and the room its true queer shape. Behold the power of love.

To recapitulate: an empty Ames room tells us nothing about perception. If properly constructed it must look like a normal rectangular room, as it gives the same image in the eye. But when objects are added the situation is different. Then the room shows that perception involves betting on odds. Perception goes wrong when common assumptions are not questioned and the true answer is unlikely. This tells us something important about the role of experience and learning, and perhaps inherited knowledge. The visual assumptions can be tested by touching the room's walls with a stick held in the hands. This gradually reduces its distorting effect on objects in the room, as slowly it comes to look its true peculiar shape (Figure 9.14).

Perhaps the Ames room demonstration is not quite so clear cut. For what happens *without the room*? If we photograph two people at different distances, from a low eye-level, do they look different sizes or different distances? There is a sort of compromise; the more distant appears both further and smaller (Gregory 1970, page 28). The Room surely adds to the size difference, but this really needs more experiments.

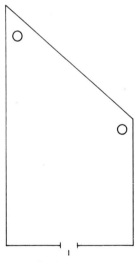

viewing point

9.14 Geometry of the room. The further wall recedes from the observer (and the camera) to the left. The figure on the left is farther away, but the walls and windows are arranged to give the same retinal images as a normal rectangular room. The figures appear the same distance but different sizes. (The nearer is about half the distance, so is twice as large at the eye.)

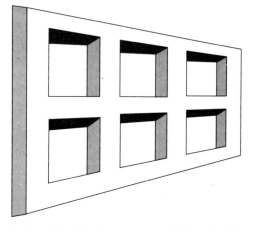

9.15 The Ames window. This is its shape, as seen at right angles. It is quite hard to see its true shape in the drawing: one does not quite know whether it is rectangular, and being viewed from the side; or whether it is a perspective-shape—which it is. When slowly rotated it goes through the most amazing transformations. Strictly, perspective is to be found only in pictures and optical images, including the eyes'—not in objects, or 'reality'. But perspective-shaped objects can upset perception of size and distance.

An equally celebrated Ames demonstration is the trapezoid window. This is a non-rectangular window-like object, made to rotate slowly by means of a motor. It is the shape of a rectangular window but as seen from a highly oblique view (Figure 9.15).

It has shadows painted upon it so that, peculiarly, as it rotates they do not change. What is seen is a complex series of weird phenomena. The direction of rotation is ambiguous, seeming to change spontaneously. This is the 'windmill' effect, observed when rotating vanes are seen against the sky, direction of rotation reversing spontaneously as near and far switch, as in the skeleton cube (Figure 10.6). A rod passing through the Ames window will sometimes seem to rotate in the wrong direction. When *its* movement is seen correctly, though the window's rotation is perceptually reversed, then the rod seems to pass through the substance of the window, like a ghost.

The window changes in size: a striking effect that is hard to describe. The point is that as the object has a marked oblique perspective shape, the normally reliable assumption of rectangular objects giving such a shape by perspective from an oblique view, misleads eye and brain into a false 'perceptual hypothesis'. These demonstrations are dramatic and eye-opening, though perhaps too complicated as research tools for measuring what is involved.

Shading and shadow

Artists make remarkably effective use of many visual cues to depth. In a pencil drawing, shading may be used to indicate the form of an object. The shading is often conventional stippling, or equally spaced lines indicating a flat region, with unequal spacing indicating that the surface is sloping or irregular.

Shading may also indicate cast shadows, which is a different matter from surface texture. Shadows indicate the direction of light falling on objects, and also that something is obstructing the light. Shadows may be caused by overhanging features, as when texture is revealed by small shadows, and then both the texture of the surface and the direction of the illumination are indicated by the form and direction of shadows. This is a matter of surprising significance. Shadows are important not only for highlighting textures for revealing the form of objects: shadow can supplement the single eye, to give something surprisingly close to binocular vision. The light-source, revealed by its shadows, replaces the missing eye of the painter.

Consider a portrait taken full face but with strong side lighting. The profile form of the nose is shown on the cheek (Figure 9.16). The

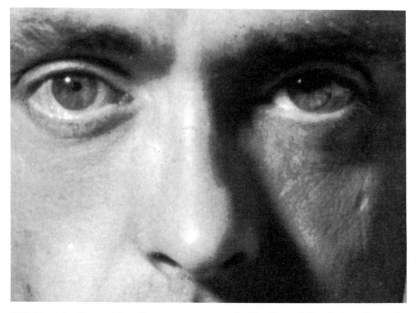

9.16 Two viewing positions from one camera angle. The form of the shadow shows the profile of the nose and eyes.

shadow thus gives us a second view of the nose. We get the same effect when looking at the moon through a telescope—indeed until space travel our knowledge of the profiles of crater walls and lunar mountains depended entirely on seeing their shadows cast by oblique sunlight. It is possible to measure the lengths of the shadows and deduce their heights and shapes. For usual object perception the visual system does this continually and it is important: the world looks quite flat when the light is behind us, for then there are no shadows, which makes for flat scenes and dull pictures.

We have already noted that depth can be reversed by optically interchanging the eyes, each eye receiving the normal view of the other (see Chapter 3). Interestingly enough, similar reversal in depth can be given by the light-source 'eye' of cast shadows being shifted from above to below (Figure 9.17). The point is, light normally falls from above. The sun cannot shine from below the horizon and artificial light is generally placed high. When, however, illumination is from below we tend to see reversed depth, much as when our eyes are switched by a pseudoscope (Figure 3.24). This effect was noted by several early writers. David Brewster (1781–1868) records it in his *Letters on natural magic*, where he describes how, when the direction of light falling upon a medal is changed from above to below, depressions become elevations and elevations depressions: intaglios becomes cameos and vice versa. This was observed at a meeting of the Royal Society in the seventeenth century, of a guinea coin viewed through a microscope. Brewster said of it:

The illusion . . . is the result of the operation of our own minds, whereby we judge the forms of bodies by the knowledge we have acquired of light and shadow.

He went on to experiment with the effect, finding it more marked in adults than in children. He noted that visual depth may become reversed even when true depth is indicated by touch. This must rank as one of the earliest psychological experiments. But almost all this was known to artists centuries before scientific experiments on how we see.

Apart from perspective, texture gradients (Figure 9.18), the hiding of parts of objects by nearer objects (Figure 9.19), haze associated with distance, and many more 'cues' are important for seeing depth and distance in the real world and in pictures.

Powers of various 'depth cues' can be measured with the technique described in Chapter 8 [page 234]. No features *determine* depth or form; they can only increase probabilities of seeing in particular ways, yet pictures can have compelling though paradoxical realism.

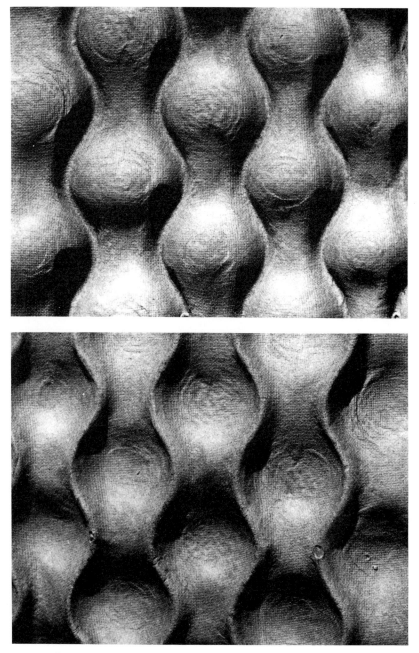

9.17 Depth-neutral shapes. (a) Illuminated from above. (b) The same lit from below. Depth reverses when the direction of illumination is changed.

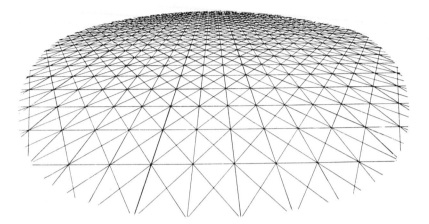

9.18 Slope from a perspective gradient. The surface texture is represented by perspective. Texture is the visible surface structure of objects, or of the ground between objects. In this example, the appearance of depth is so powerful that it may seem 'direct', not requiring interpretation. But as for any other picture, this is open to alternative interpretations. Indeed, it is actually flat!

Realities of *trompe l'oeil*

Occasionally, especially in architecture, the artist goes all out to create the most compelling possible illusion of reality. Illusions may fill rooms with statues, or comple missing parts of churches—with the superb technique of *trompe l'oeil*. With its absence of brush strokes or other surface texture, this tradition goes back to ancient Greek painting—when it was claimed that bees could be fooled by painted flowers. Yet admiration is barbed with criticism that 'this is not quite art'. Thus, Lucas de Here said of Van Eyck's almost too realistic pictures: 'They are mirrors, yes, mirrors, not paintings.'

When depth is shown very realistically, depicted objects rotate—following every movement of the observer. This can be disturbing. Just why this dramatic movement is seen in flat perspective pictures, that lack surface texture is not entirely clear. Easy to achieve by stereoscopic projection, this is an interesting phenomenon which needs more investigation. The seen movement is surprising and perceptually revealing, because it does *not* depend on changes at the eyes. For if the *picture* is rotated nothing happens. This is worth trying. The seen motion depends upon the observer's hypothesis of his or her body movement through space. We know too little about the subtle relations between seeing and doing.

9.19 A cunning trick of occlusion for upsetting depth perception. The two sets of squares, though they appear in the same order, are differently arranged in distance. In the third row of playing cards the Jack is nearer than the six. This works because occlusion is a powerful visual cue; here it is misleading because what should be occluding parts of nearer cards are cut-outs.

No doubt this apparent motion of *trompe l'oeil* would be too disturbing for seeing pictures in a gallery. Artists gain more by settling for less realism than is technically possible.

10

Illusions

What are illusions?

When a perception departs from the external world, to disagree with physical reality, we say we experience an illusion. So an illusion differs from truth. But how do we know the truth? There are different accounts of truth through the history of science—and these differ from accounts in art, religion, and in the many flavours of mysticism and metaphysics.

There are accepted realities of common sense by which we live from day to day, but when described explicitly these turn out to be riddled with contradictions. Contradictions of common sense and appearances feed philosophy and science, to generate counter-intuitive claims to reality such as quantum physics and relativity, which are now accepted as scientific truth. But these accounts are so different from how things appear that to take them as reference would force us to say that all perceptions are illusion—which is not helpful.

As naïve appearances have supported survival for a long time in a great variety of situations, we should not dismiss them as entirely false. But, for example, it doesn't matter for survival whether the sun moves or the Earth rotates; but this does matter for whether the appearance is an illusion.

It turns out to be extremely hard to define 'illusion'. To say that *all* appearances of objects are illusions is no more helpful than to say that all experience is a dream. Although logically irrefutable, this drains useful meaning from 'dream' and from 'illusion'. (Similarly, there is little point in saying that everything is beautiful or everything ugly: perception and language need contrasts to have meaning.) We think of *objects* as real. But what is an object?

What is accepted as an *object* depends largely on use—such as tables for putting things on and chairs for sitting on. A book on a shelf is an object; but when we are reading, each page or word or letter becomes an object. Object classification is highly dependent upon behaviour. This may suggest a useful distinction between 'illusion' and 'reality'. For we might say that illusions are departures from the world accepted as real for behaviour. On this account they are not departures from recondite physics; but rather from behaviourally accepted objects with characteristics such as lengths and angles and curvatures, which are measured very simply. So illusions can be assessed by comparing what is experienced with what is simply measured with rulers, protractors, scales for weight, clocks, and so on. We can use measurements of carpentry and cooking for measuring illusions of perception.

Some phenomenal phenomena, however, cannot be matched against accepted reality or simply measured. For there are perceptual *fictions*, which have no object counterpart. Others which are hard to match or measure are perceptual *paradoxes*, which cannot exist as objects. Others flip from one object perception to another as alternative perceptual *ambiguities*, which only occasionally could match object reality.

The frequently used term 'optical illusion' is misleading when the cause is beyond the eyes' images. Although this is so for most illusions, some can appropriately be called 'optical', or more generally 'physical' from their cause. Perhaps the most familiar example of an optical illusion is mirror images. Seeing oneself double—with a silent touchless extra-self in a wrong place, through the looking glass—is deeply paradoxical though resulting from quite simple principles of optics. The cause is physical; but it is also due to the visual brain being unable to cope with the situation. Yet this remains perceptually paradoxical though we understand, conceptually, what is going on. If understanding of reflections fed into visual processing, mirror paradoxes might disappear. So though this cause is physical, the appearance depends also on lack of visual cognition. It can be hard to decide which phenomena of illusion are *physical* and which are *psychological*, and of course *physiology* is always active in perception. There are often mixed causes. If we call a mirror paradox 'physical', this attributes its primary cause to optics.

When Sir Isaac Newton walked under a rainbow to his house, surprisingly the house got nearer and nearer, until he went through the door; but the rainbow got no nearer (Figure 10.1). Newton could not walk under its arch of colours as he could through the doorway; yet his perceptual apparatus of eye and brain was working fine for both. It is when a rainbow sets up expectations of normal touchable objects that it misleads, to be an illusion.

Sir Isaac Newton and other physicists, including René Descartes,

10.1 Newton's house (at Woolsthorpe, near Grantham in Lincolnshire) with a rainbow. Why is the rainbow an illusion though the house is real?

realized hundreds of years ago that rainbows are within physics—and explained by laws of optics (Figure 10.2). They would not be deluded into seeing it as like a normal object, such as a house; so for them perhaps the rainbow is not illusory. This raises tricky issues of what are *perceptual* and what are *conceptual* illusions.

The fact that perceptions can depart from physically accepted

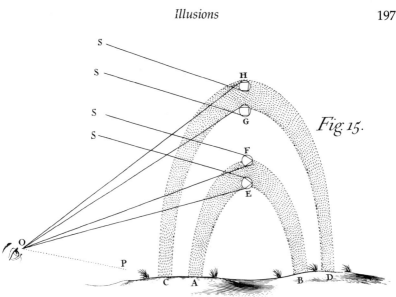

10.2 Newton's drawing of the optics of a rainbow. Sunlight is reflected once and refracted twice in each drop. From each, a single colour emerges—depending on the height of the drop—many drops at various angles producing the spectral colours spread across the sky. (From *Opticks*, 1704)

realities of objects has philosophical implications and practical consequences. It tells us that our perceptions are not always, and very likely never, directly related to physical reality. In practical terms it shows that we not only 'take our lives in our hands'; we trust betting on the odds of eye and brain.

This is unfamiliar country, but for several reasons it is worth taking our journey towards explaining illusions. As for the natural sciences, it turns out to be important to organize the phenomena into categories. This certainly paid off for chemistry and biology, and for kinds of stars. As for these, we need to consider appearances and underlying causes. Phenomenal phenomena are weird experiences, which become even more intriguing as we come to understand them. Here art and science meet—eye to eye.

Some illusions can cross between the senses. Best known is the size–weight illusion: a *smaller* object feels heavier than a *larger* object of the same weight. This is worth setting up yourself—to experience literally at first hand (see box).

Although physics and physiology are involved when a weight is lifted, the reason for this powerful illusion is not within the concepts of either: it is a *cognitive* illusion. For it depends on one's *knowledge* of

---

Seen size affects felt weight

Try filling a small can and a larger one, with, for example, sugar, to the same weight. Pick them up one at a time, or simultaneously with both hands. The smaller will feel considerably heavier. Try it blind-folded. When there is no evidence for the size difference, the size–weight illusion disappears.

The muscles are set for *expected* weights. As larger objects are usually heavier than smaller objects, the smaller weight calls for less muscle force—so it seems surprisingly heavier than the larger weight.

---

objects—that larger objects are generally heavier than smaller objects of a similar kind. The illusion is surprising. Although it hardly fits physics, it has a rational explanation which tells us a lot about perception.

Perception can go wrong in many ways. Most dramatically, entire inner worlds may be created as seen fictions are mistaken for reality. This can happen in drug-induced states or in mental illness, especially schizophrenia. In addition to hallucinations where experience departs altogether from reality, we all see certain things or patterns as distorted—what are misleadingly called 'optical illusions', though the cause is in eye or brain. But let's start with:

Dreams and hallucinations

Both dreams and hallucinations may be visual or auditory, and may involve any of the other senses, such as smell or touch or pain. When several senses are combined the impression of reality can be over-whelming. Evidently hallucinations can be socially determined, for there are well-attested cases through history of many people together experiencing events which never occurred.

The Canadian brain surgeon Wilder Penfield (1891–1976) triggered memories and produced hallucinations on demand by stimulating the human brain with weak electric currents. Brain tumours may give persistent visual or auditory experiences; the 'aura' preceding epileptic seizures may also be associated with hallucinations of various kinds. In these cases, the perceptual system is moved to activity not by the normal signals from sensory receptors, but by brain stimulation. In Penfield's experiments, these patients undergoing surgery generally experienced something of what was going on in the operating theatre at the time, together with the hallucinations.

Drugs can entirely replace sensed reality with hallucination. It remains somewhat mysterious how hallucinogenic drugs, such as LSD (D-lysergic acid diethylamide) affect vision, but they seem to stimulate regions of brain active in normal perception. These drugs may affect other species in various ways. They disrupt the web spinning of the spider *Aranea diamata* so that it forms abnormal but characteristic patterns, and monkeys reach out for non-existent objects.

In us, imagery similar to that induced by drugs can occur in half-waking states (hypnagogic dreaming) when the experience can be like a Technicolor film, vivid scenes apparently passing before the eyes although they are shut. In this lucid dreaming the dreamer may have some control. This is more like Virtual Reality than the cinema. Hallucinations have also been found to occur when people are isolated in solitary confinement in prison, or experimentally in isolation chambers, in which the light is diffused with special goggles and nothing happens for hours or days on end. It seems that in the absence of sensory stimulation the brain can run wild to produce fantasies. This is what happens in schizophrenia, when the outside world makes little contact with the individual, so that he or she is effectively isolated. These effects of isolation are interesting not only from the clinical point of view; they offer hazards in normal life. Workers may be effectively isolated for hours with very little to do, in industrial situations where responsibility is taken from the operator by automatic machines needing attention only occasionally. In single-handed sailing and space flights prolonged isolation occurs. The hazards are sufficient reason for sending more than one person on each mission into space.

Throughout history, dreams have been regarded as significant; the Greeks had special houses to be visited for incubating dreams. Dreams have always excited wonder and often more—affecting personal and political decisions, sometimes with terrible results.

For the mystic, dreams and hallucinations are insights into another world. For them the brain is seen as a hindrance to understanding; a filter between us and a supra-physical reality, seen only when the brain's normal functioning is disturbed by drugs, disease, or starvation. For the more down-to-earth empiricist philosophers the brain is to be trusted only in health. Although interesting and perhaps sometimes suggestive, for science hallucinations are seen as aberrant outputs of the brain, to be mistrusted and even feared, though of course Freud is an exception here. Aldous Huxley (in *Doors of perception* 1954) describes most vividly the viewpoint of the mystic. But probably all neurologists and most philosophers hold that drugs distort and impair perception and thinking, with a danger of producing long-term effects, without revealing truths from any outside source. The general question is whether there is any extra-sensory perception. This writer, not finding convincing evidence in spite of its universal historical acceptance, takes it to be a sad delusion.

Although there is no evidence, in the author's opinion, that hallucinations convey new knowledge, no doubt dreams, and other brain activity free of sensory control, may show something of hidden motives and hopes and fears, revealing something of the hidden individual. This, however, is not a main interest here.

If the general view taken here is correct—that perceptions of objects are psychological projections of stimulus patterns, and projections of knowledge and assumptions into assumed reality—the status of perception for knowledge is not as secure as empiricists would like. For perceptions are not *essentially* different from hallucinations. This is a major reason for transcending individual perceptions with shared observations, and with the controlled experiments of instrument-aided science. How does this fit with truth in art?

Op art—and all that jazz

There are pictures which are extremely disturbing—jazzing and moving, and generating ghostly shapes and colours. These are part of Op-Art, best known in the work of Bridget Riley (Figure 10.3). The cause may be mundane, but does this reduce the significance of the experience?

Some of these remarkable effects are seen in simple patterns of repeated parallel lines or converging rays (Figures 10.4 and 10.5), which were studied by the visual scientist Donald MacKay in the 1950s.

Moiré patterns

If two identical overhead projector transparencies of MacKay's patterns are sandwiched, and if one is moved slightly across the other, exactly these patterns are seen, projected on the screen. This is evidence that the (single) ray figure is producing moirés—by 'beating' with retinal after-images, following small eye movements.

The jazzing occurs in these figures, and in Op-Art, even though the eyes are held as still as possible. It is known that the eyes' lenses are in

10.3 *Fall* by Bridget Riley. In this remarkably powerful picture, eye movements and 'hunting' of the lens of the eyes shift the retinal image across the receptors, and produce 'beating' with each momentary after-image, produced especially when the eyes pause between their sudden jumps (saccades). At least, this is one theory. The physiologist Semir Zeki thinks very differently that movement centres of the brain (in area V) are directly stimulated by these patterns to give experience of motion. So we have a controversy.

continual motion, as they hunt for sharp focus (accommodation), and the lens has some dynamic astigmatism shifting the image very slightly. As the repeated lines must be quite closely spaced, the image may shift from dark to light and back again at each border—stimulating the retina's on–off receptors, associated with the normal signalling of movement. There are, however, authorities who believe these effects arise directly from movement systems in the brain rather than from the eyes. It should be possible to test between these ideas by stabilizing the image on the retina; or even better, by by-passing

10.4 Mackay rays. This jazzes about dramatically, and ghostly chrysanthemum shapes appear. These seem to be moiré patterns, the rays 'beating' with their momentary after-images, with small eye movements, or displacements from hunting of the lens for accommodation. Some authorities, however, believe the effects to be directly brain induced, without motion signals from the eyes.

10.5 Closely spaced parallel lines have similar effects to the ray figure. These are the basis of many Op-Art pictures, used by Bridget Riley and other artists to stunning effect.

fluctuations of the lens—which can be done easily with a flash after-image (see box).

De-jazzing with an after-image

If the Ray figure is viewed in complete darkness, illuminated for a small fraction of a second with an electronic camera flash, a vivid after-image will be produced, fixed precisely on the retina. Does this show the jazzing and the chrysanthemum patterns normally seen in the Ray figure? If not, a brain-induced explanation is pretty well ruled out, in favour of movement signals from the eyes, due to 'hunting' of the lens for accommodation.

For this experiment, it is essential for the eyes to be directed at the centre of the figure, with appropriate accommodation for its distance—but a 'fixation' spot of light must not illuminate the figure before or after the flash. The eyes can be accommodated to the correct distance by placing a sheet of glass inclined at 45° and placing a small spot of light to the side, at the same distance as the figure, to reflect from the glass.

What happens?

If you try this experiment, you will see that the jazzing effects disappear. If so, they may be attributed to movement mechanisms stimulated by motion of the repeated lines of these figures.

Classifying phenomenal phenomena

It is important to classify phenomena in science. So we need at least tentative classes of illusions with names to identify them. The classes may depend on appearances and on causes. But as causes of many illusions are unknown, or controversial, and appearances of some can be hard to describe, this is not a simple project and we cannot expect to achieve complete agreement or immediate success.

It is suggestive that errors of language neatly match kinds of visual illusions. So we may start by using the names of language errors to name appearance categories of illusions. We will call them *ambiguities*, *distortions*, *paradoxes*, and *fictions*.

It may be no accident that these correspond both to errors of language and to errors of perception. Both perception and language give descriptions, and both depend upon how objects and situations are classified for behaviour. It may indeed be that very ancient, pre-human, perceptual classifications are the basis of the structure of languages. Perhaps language developed so fast in humans because it built upon the perceptual experience of many millions of years of classifying objects and actions giving nouns and verbs.

Let's look at these four kinds of illusions—ambiguities, distortions, paradoxes, fictions—and spell out the differences:

Ambiguities

Any retinal image is infinitely ambiguous. That is: it could correspond to, or represent, an *infinity* of possibilities of shapes and sizes and distances of objects. This is easily seen with an ellipse: it could be given by a tilted circle, a small near circle (or ellipse), or a farther correspondingly larger circle or ellipse (Figures 9.8, 9.12). The possibilities are endless—yet amazingly, we generally see just one of the infinite range of what might be out there. How this is done remains an only partly solved puzzle of perception.

There are well-known figures which flip between a few possibilities. These are known as 'ambiguous figures'. They are extremely important for showing the dynamics of perception, the searching for hypotheses of objects that might or might not be in the external world. Here the answer—the perception—is never decided. There are different kinds of flipping ambiguities—between shapes, depths, and different objects. The best known is the Necker cube (Figure 10.6).

Here there is no evidence to indicate which of the large faces is the front or the back. Vision entertains alternative, roughly equally likely hypotheses. So here we see it flip between two equally likely cubes as different depth hypotheses are entertained. What is not clear, is why it

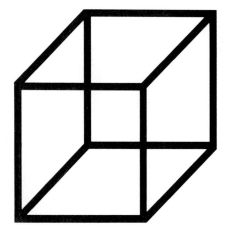

10.6 Necker cube. This flips in depth. Discovered by a Swiss crystallographer, L. A. Necker in 1832, while drawing rhomboid crystals seen with a microscope. Suddenly the drawing (or the crystal) flipped so they looked quite different!

is only these depth hypotheses that are entertained and seen. Some well-known change-of-object ambiguities are illustrated in Figures 10.7 and 10.8.

Ambiguities can be extremely useful for perceptual research: as perceptions change though the input to the eyes remains unchanged—so we can see what is going on 'from the inside'. In particular, visual ambiguities allow us to separate effects of bottom-up signals from the eyes, from top-down knowledge and assumptions (Figure 10.9). This is a very important distinction. There is a wide range of opinions on their relative importance. The more the top-down contribution, the less 'direct' is perception. This is bad news for empiricists seeking certainty for seeing, but good news for those who think of perception as intelligently creative—making effective use of limited available data to represent what might be out there.

A beautiful demonstration by the Austrian physicist Ernst Mach, (1838–1916), who was extremely interested in visual perception and made interesting contributions, is the Mach corner (Figure 10.10). This is strong evidence for top-down knowledge and assumption affecting bottom-up signalling of brightness. The general implication is that all 'raw data' are cooked!

10.7 Old wife or young mistress? E. G. Boring's object-ambiguous figure. By looking at different features (ear, neck, nose), this may be flipped at will between the two likely alternatives.

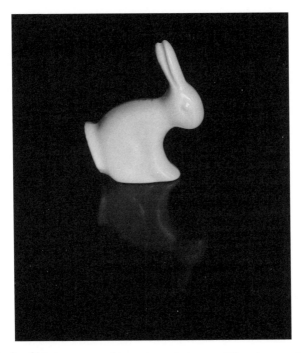

10.8 The duck–rabbit is an even bigger jump between different perceptual object hypotheses–perceptions. (If the figure is rotated by 90°, one will be preferred over the other, as it becomes more likely.)

The hollow face

There is a general rule, or visual assumption, that objects are likely to be convex rather than hollow. Thus, looking into the inside of a box, with one eye, it will generally switch in depth to appear as the outside, though one knows one is looking into the inside. When the box is held in the hand and rotated gently, the effects of reversed movement are bizarre. (Fascinating variations on this theme have been invented by the illusionist Gerry Anders.)

The most dramatic demonstration that probabilities of specific objects are important, is the *hollow face*. This may be a mask of a normal face on its outside, and hollow on the inside. (These are available in joke shops.) From a metre or so, the inside face will appear as a normal, nose-sticking-out face, though it is hollow. This works, viewed with both eyes in normal lighting. If the mask is slowly rotated it goes through astonishing transformations, reversing direction as the hollow back or the sticking-out front appears, much as for a

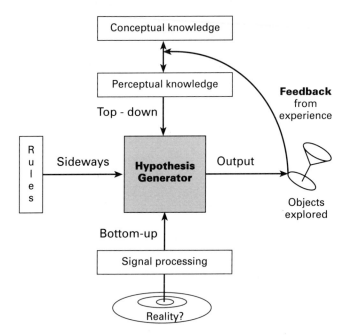

10.9 Ins and outs of vision. Not to be taken too literally, this is a way of relating the notions of *bottom-up* signals from the eyes, *top-down* knowledge, and what we are calling *sideways* rules of perception, such as the Gestalt laws of organization and perspective. This is developed in Chapter 11, 'Speculations', with an attempt to classify illusions.

wire cube when back and front reverse, though more dramatically (Figure 10.11; and see also Plate 3).

What this demonstrates is important: top-down object knowledge can dominate bottom-up signals.

The more realistic and typical the face, the better it works— remaining apparently normal down to a nearer viewing distance. When upside down, it works less well, as it is now less typical of a face. These simple experiments are well worth trying. They must be seen to be believed.

Distortions

Apparently simple sensations, even brightness, can be distorted. Figure 5.3 is a nice example. Lengths may be distorted, and a straight line may be bent into a curve, so it is difficult to believe it is really straight. Virtually everyone sees these distortions similarly. Many were discovered by physicists and astronomers in the 19th century, when designing hair-lines for eye pieces of optical instruments—to make

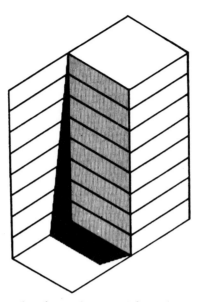

10.10 Mach corner (or card). A shadow (or painted shadow) on a card bent to this shape, changes brightness, or darkness, each time the corner flips in depth. When 'in' it is lighter than when 'out'. This is because when it is seen as a shadow (more likely when 'in'), it is minimized visually, as shadows are not important objects.

their measurements more accurate! Several bear the names of distinguished German optical-physicists concerned with avoiding observational errors, but gaining immortality through named illusions.

There are several causes of distortions, and it can be hard to establish whether a distortion is physiological or cognitive in origin. One way of finding out is to discover what destroys the illusion. For several reasons we believe the distortion of the *café wall* illusion to be disturbances of retinal signals of edges—quite simply physiological.

The café wall illusion

This was discovered in Bristol on the tiles of a 19th century café—hence its name. The long illusory wedges seen in Figure 10.12 at first sight contradict the notion that distortion illusions are due to inappropriate size scaling, set by depth cues such as perspective (page 175)—for here there is no perspective, or other depth cues. Is this example evidence against the theory—or is it a different kind of illusion? This shows the importance of classifying phenomena. Without classifications we don't know how to apply evidence, or assemble evidence for confirming or attacking hypotheses.

10.11 Hollow mask. Though this is a hollow mask, it appears to be a normal nose-sticking-out face—because a hollow face is so unlikely. When rotated, it is seen as turning in the wrong direction. This is because motion parallax is wrongly interpreted. This works with both eyes open, down to a metre or so, when (bottom-up) stereo information beats (top-down) knowledge of faces. It continues to work though one know this face is hollow. (The relative powers of bottom-up signals and top-down knowledge can be assessed by measuring the viewing distance of the change over.)

It turns out that models of the café wall, allowing many variables to be introduced, show that this has a rich lawful set of phenomena.

We should start by noting that the long illusory wedges are not due to errors in the positioning of the 'tiles' horizontally (which would be very boring). If alternate rows are slid across, by half a tile width, the direction of the wedges reverses. Nothing else reverses them.

10.12 The café wall illusion. This has large systematic distortions—the long wedges—yet here there is no perspective. It has only parallel lines and lines at right angles. So what could cause this massive distortion?

It turns out that this illusion is highly lawful (see box).

Laws of café wall distortion

The wedges reverse when alternate rows of tiles are displaced by half a square (or 'tile') width.
The distortion is lost when:

- The 'mortar' is *brighter* than the light 'tiles' or *darker* than the dark 'tiles'.

- The 'mortar' is thick—subtending more than 10' arc to the eyes.

- The tiles are different colours (for example, red and green) with the *same brightnesses* (equi-luminance).

The café wall seems to flout a basic feature of symmetry—the Curie principle. The French physicist Pierre Curie (1859–1906), stated that *symmetry cannot produce asymmetry*. But this figure is symmetrical at the small scale (as any region of displaced tiles is the same as any other region) yet it produces large-scale asymmetry. What is going on?

It turns out that there are two processes. The first is small scale: where there is brightness contrast across the neutral mortar line, half the dark and light tiles move towards each other—forming small scale wedges where there is local asymmetry. The eye integrates these little wedges into the long wedges that are seen.

Perspective illusions

Well-known distortion illusions such as those in Figure 10.13 are very different. They obey completely different laws, and they are destroyed by entirely different conditions. They are all perspective drawings of typical scenes or objects. (Here I have given the game away by calling them 'perspective' illusions, but is this the right game?)

Physicists and physiologists, psychologists and philosophers, have tried to explain these distortion illusions for over a hundred years. Current explanations are controversial; but I believe we can develop an explanation for these distortions, which throws light on the nature of object perception. But first, let's look at various theories that have been proposed. They reflect different views of perception itself, and are remarkably varied.

Theories of 'perspective' distortions

The pregnance or 'good-figure' theory

The idea of 'pregnance' is central to the German school of Gestalt psychologists' account of perception. The English meaning is similar

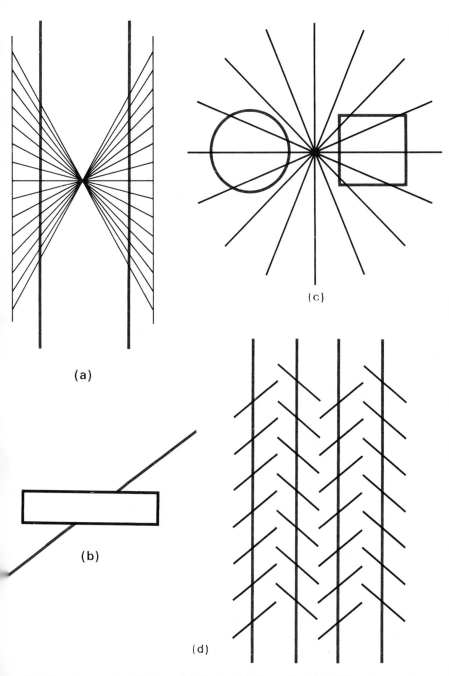

10.13 Four 'perspective' illusions: (a) Hering, (b) Poggendorf, (c) Orbison (two combined), (d) Zöllner.

to the use of the word as in 'a pregnant sentence'. The illusions are supposed to be due to pregnance exaggerating the distance of features that seem to stand apart, and reducing distances where features seem to belong together.

The status of the notion of pregnance is doubtful. Certainly random or systematic arrangements of dots do tend to be grouped, in various ways, so that some 'belong' together while others are rejected or are organized into other patterns (Figure 1.1) but there seems no tendency for the dots to change their *positions* as a result of such grouping; though surely this should be a prediction of the pregnance theory of distortion.

Empathy theories

It was suggested by Theodore Lipps (based on an idea of the American psychologist R. H. Woodworth) that the observer identifies himself with parts of the figure (or the pillars of a building) and becomes emotionally involved—so that vision is distorted rather as emotion may distort an intellectual judgement. In the Muller–Lyer illusion, (Figure 10.16, page 219) it would be argued that the outgoing 'arrows' suggest—emotionally—expansion, which one then sees.

It is true that a very thick column supporting a narrow cornice looks clumsy. And perhaps one does imaginatively stand in the place of the column—as Hercules took the load of the sky off the shoulders of Atlas before turning him to stone. The caryatids of Greek temples (Figure 10.14) embody (quite literally) this idea in architecture. But although of immediate relevance to aesthetics, it can hardly be taken seriously as a theory of the distortion illusions. The Muller–Lyer arrow figures (for example) give distortion whatever one's mood, and continue to do so when any initial emotional response would surely have died through boredom. There may be perceptual effects of strong emotion, but the illusion figures would seem singularly devoid of emotional content—except to those who try to explain them! More serious: the distortions are virtually the same for all observers though emotions are generally different and change from time to time.

Eye movement theories

These suppose that features producing the distortion make the eyes look in a 'wrong' place. In the arrow (Muller–Lyer) illusion, it may be supposed that the eyes are drawn *past* the lines by the arrow-heads, making the lines look the wrong length. The alternative version is that

10.14 Man enough for the job? Perhaps we identify ourselves with pillars, so that there is a right size in human terms for carrying the load. Is this the basis for distortion illusions? (Caryatids of the Erechtheion, Athens, 421–405 BC, designed by Mnesicles.)

the eyes are *drawn within* the lines. But it is clear that neither can be correct. The retinal image can be optically fixed on the retina (Figure 3.17), or more simply (as the reader can easily try) by illuminating the figures with a photographic flash and viewing the after-image, which is stuck on the retinas and moves precisely with the eyes. The result?—the distortions remain unchanged.

Also, the distortion goes in opposite directions for the pair of arrow figures—but the eyes can't possibly move in different directions at the same time.

The eye movement theory is sometimes stated in a different form: that it is not *actual* eye movements, but *tendencies* to make the eyes move which produce the distortions. But as the eyes can't move in different directions at the same time, presumably this is also impossible for tendencies or intentions to move the eyes. There is no evidence for the eye movement theories and plenty of evidence against them.

Limited acuity theories

Considering the Muller–Lyer, or arrows illusion: we might well expect the figure with the outgoing fins to look too long and the figure with the ingoing fins to look too short, if the acuity of the eyes was so low that the corners could not be clearly seen—or optically resolved. This may be demonstrated by placing a sheet of tracing paper over the figures, when a slight change of length might appear. The effect is, however, far too small to explain the Muller–Lyer illusion, and does not apply to many of the other figures; so at least in my view this is not a serious candidate. Some authorities do, however, hold similar ideas, generally couched in terms of low spatial-frequency visual filters. These distortions remain, however, when (by careful, clever means) low spatial-frequency components are removed from the figures. I reject this kind of account for these figures; but retinal lateral inhibition is important for some figures depending on brightness differences, such as the café wall illusion (Figure 10.12).

Feature detectors theories

Could it be that these figures upset the orientation detectors (page 77) discovered by Hubel and Wiesel? We may distinguish two kinds of theory along these lines:

1 That orientation detectors exaggerate acute angles, and minimize obtuse angles. This generalization was suggested by Helmholtz,

revived since; but there is no physiological evidence for such a physiological cause.

2 That there are interactive effects between orientation (or angle) detectors, perhaps like lateral inhibition interactions in the retinas of the eyes.

It may seem unlikely that the initial stages of pattern detection would be so 'ill-designed'; though, as Colin Blakemore has pointed out, for retinal interactions, they may be an inevitable side-effect of generally useful *lateral inhibition*. This is interaction between cone receptor cells such that a strongly stimulated cone cell inhibits the activity of nearby cells, giving a kind of sharpening effect on the signalling of edges where there is blur. Blakemore's suggestion is that converging lines produce lop-sided lateral inhibition, shifting the neurally signalled peaks, to produce distortions.

Evidence against this as a major effect, is that distortions still occur when the figures have colour contrast but no brightness contrast (isoluminance, or equi-luminance), though it is believed that lateral inhibition only works when there are luminance differences. Also, the distortions still occur when different parts are presented to each eye, by Julesz's technique (Plate 2) and combined by stereoscopic fusion. Here the lateral inhibition could not be retinal.

Some visual illusions do have their origin in the retina—before brain or mind. The technique invented by Bela Julesz for sharing pictures between the eyes, with patterns of random dots, can be used to see whether a distortion illusion originates in the retinas or further up in the brain, after the signals from the two eyes are combined. If the figure is only visible by the eyes combining their (dotty) signals, and the distortion is seen, it must originate in the brain, not in the eyes.

This does not reject *cortical* interactions, which remain a possibility. There seem, however, to be better alternatives. This takes us to very different cognitive accounts, including:

Perspective theories

The central idea is that many of the distortion figures suggest depth by perspective, and this produces size changes.

Consider the illusion figures we started with (Figure 10.13). These can be fitted very naturally to typical perspective views of objects lying in three-dimensional space. These figures are flat projections of three dimensions.

Further, it is true of these figures that features represented as *distant* appear *expanded*. This is a promising starting point. It immediately

10.15 Ponzo illusion. This is a simple perspective figure. The (upper) rectangle, represented as more distant, is expanded.

suggests that these phenomena are cognitive rather than simply physiological; that they are associated with rules and knowledge of mind rather than to disturbance of physiological function of eye or brain.

The phenomena of represented distance producing expansion is clearest in the Ponzo figure (Figure 10.15). The converging lines are typical depth perspective: placing the upper rectangle distant somehow expands it.

The most famous distortion illusion of all is the Muller–Lyer 'arrows' (Figure 10.16). These are typical perspective drawings of corners. The outgoing fins are two-dimensional projections, say of the inside corner of a room—the ingoing arrow-heads are perspective drawings of the outside corner of a box or building (Figure 10.17).

These illusion figures are typical perspective views of scenes or of objects in depth. (We may note, however, that in all cases they *could* be drawings of something quite different, though atypical. This is so, as any retinal image is infinitely ambiguous, though generally but one interpretation is seen.)

Why should perspective depth produce distortion in pictures? The traditional perspective theory simply states that these figures 'suggest' depth, and if this suggestion is 'followed up', the most distant features appear larger. But how could 'suggestion of distance' produce changes of apparent size? Further, why should suggestion of greater distance produce *increase* in size—though distant objects, giving smaller retinal images, are normally seen somewhat *smaller*? The traditional perspective theory fails to provide a *modus operandi* for the distortions, and its predicted size changes are the wrong way round.

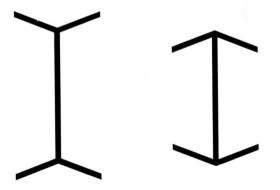

10.16 Muller–Lyer illusion. The line with the outgoing fins looks considerably longer than the equal length line with the ingoing fins. This illusion has received more experimental treatment than any other, perhaps because it is simple, clear-cut, and easy to measure.

10.17 The Muller–Lyer figures can be seen as flat three dimensional drawings of corners.

Although these predictions of the perspective theory go exactly the wrong way round, this is far better than predictions unrelated to facts. We will try to develop a theory along perspective lines which leads to correct predictions. It is worth devoting some time to this, as by establishing connections between phenomena we gain more general understanding of perception.

There is a perceptual mechanism which is capable of producing such distortions. This is size-constancy scaling. Visual scaling compensates for changes of retinal image sizes, with changes of object distance—so things do not shrink with increasing distance, nearly as much as their images get smaller (Figure 10.18). Consider an audience at a theatre—the faces all look much the same size, yet the retinal images of the further faces are far smaller than the nearer.

This is a most intriguing process, which we can see operating in ourselves. It can go wrong. When it does, instead of keeping the scale of things almost constant in spite of changes of object distance, it can produce a variety of distortions. The box describes a powerful demonstration.

Are your hands the same size?

Look at your two hands, one placed at arm's length, the other half this distance, at the elbow. They will look almost exactly the same size; yet the image of the further hand will be only half the (linear) size of the nearer. But if the nearer hand is brought to *overlap* the farther, then they *will* look very different in sizes. The overlap defeats constancy scaling, showing what perception would be like without it.

This little experiment is well worth carrying out. It may be done also with unknown objects, such as pencils which may be different lengths. Scaling does not depend on knowledge of sizes of the objects.

What we are calling size constancy was described more than three centuries ago by René Descartes (Figure 10.19), in his *Dioptrics*, of 1637:

I need not, in conclusion, say anything special about the way we see the size and shape of objects; it is completely determined by the way we see the distance and position of their parts. Thus, their size is judged according to our knowledge or opinion as to their distance, in conjunction with the size of the images that they impress on the back of the eye. It is not the absolute size of the images that counts. Clearly they are a hundred times bigger [in area] when the objects are very close to us than when they are ten times farther away; but they do not make us see the objects a hundred times bigger; on the contrary, they seem almost the same size, at any rate so long as we are not deceived by (too great) a distance.

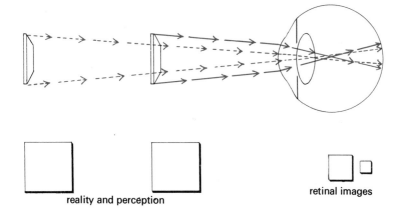

reality and perception

retinal images

10.18 Constancy scaling. The image of an object halves in size with each doubling of distance of the object. But it does not *appear* to shrink anything like so much. The brain compensates for the smaller image with increased distance, by size-constancy scaling.

Here is as clear a statement of size constancy as any made later by psychologists; except that Descartes greatly over-states the importance of knowing the sizes or shapes of the objects. He goes on to describe what is now called shape constancy:

Again, our judgements of shape clearly come from our knowledge, or opinion, as to the position of the various parts of the objects and not in accordance with the pictures in the eye; for these pictures normally contain ovals and diamonds when they cause us to see circles and squares.

It is odd that this was not taken up until quite recently, very likely because it did not fit prevailing ideas of physiology or philosophy.

The ability of the perceptual system to compensate for changing distance was investigated in the 1930s by the English psychologist Robert Thouless. Thouless measured size and shape constancy under various conditions for different types of people, using very simple apparatus—rulers and pieces of cardboard. For measuring size constancy, he placed a square of cardboard (which unlike, say, a hand could be of any size), at a given distance from the observer, and also a series of different-sized squares at a nearer position. The subjects were asked to choose the nearer square which appeared the same size as the further square. From the measured sizes the amount of size constancy could be calculated. Thouless's subjects generally chose a near square of almost the same size as the distant square, though the eyes' images were very different sizes. Constancy was generally almost perfect for fairly near objects, though it broke down for distant objects, which can look like toys.

10.19 René Descartes (1596–1650), perhaps the most influential of modern philosophers. It is now difficult to escape from his duality of mind and matter, which permeates almost all modern thought in psychology. He clearly described perceptual size and shape constancy, long before they were studied experimentally.

Thouless also measured shape constancy, by cutting out a series of cardboard ellipses, of various eccentricities, which were selected by the subject to match a tilted cardboard circle. Again, it was found that constancy could be nearly perfect—depending on the depth cues available. Thouless's 'phenomenal regression to the real object' might be taken to imply that we have some *other* knowledge of object reality. This phrase is no longer used. There has been serious confusion over constancy as a *process* and constancy as the *result* of visual processes. So the term 'size scaling', for the process producing size constancy is useful; and we may also speak of 'shape constancy' by scaling.

It is possible to see the mechanism of one's own size scaling at work with a simple demonstration of what is called Emmert's law for after-images (see box).

Emmert's Law

First, obtain a good clear after-image, by looking steadily at a bright light, or preferably a photographic flash. Then look at a distant screen or wall. The after-image will appear to lie on the screen. Then look at a nearer screen. The after-image will now appear correspondingly nearer—and will look smaller. With a hand-held screen (such as a book) moved backwards and forwards, the after-images will expand as the hand moves away and shrink as it approaches the eyes, though of course the retinal after-image remains of constant size. We therefore see the brain's *scaling* changing as the distance of the screen changes. The after-image is seen to (nearly) double in size with each doubling in distance of the screen. This is Emmert's law. It is typical of the psychological projection of retinal images into external space.

Clearly, we are seeing size scaling changes which would normally *compensate* changes of retinal images with object distance. For any object, there is only one correct perceived size. This is normally given when the distance is correctly perceived. Can scaling be set *inappropriately*—to produce distortion illusions? Is this the key to an adequate theory of perspective distortion illusions?

Inappropriate constancy-scaling theory

It is one thing to have a theory, quite another to prove it; or at least to show that it is the best of available suggestions. If these distortions are always related to distance, it may indeed seem reasonable to think that perspective sets size scaling—to give errors of perceived size when distance is seen incorrectly. This looks reasonable; but there is a difficulty, which blocked development of a theory along these lines for many years. The difficulty is that the illusion figures generally appear *flat*. If constancy always follows perceived depth (as in Emmert's law for after-images), then clearly there is something seriously wrong (or missing) from this account. We shall take the view that the notion is essentially correct, but not complete. Can we complete it?

The reason why the illusion figures appear flat, in spite of their perspective depth features, is not difficult to discover. In the first place, if they are shown on normal textured paper there is visual evidence that they are indeed flat. All perspective pictures have a curious depth paradox: they *represent* depth, with their perspective and other depth cues; yet as *objects* the pictures *are* flat and their textured surfaces provide depth cues showing that they are flat.

We can, however, remove these cues—by making the surface invisible. This can be done for example by painting or drawing with luminous paint. They may be made to glow with soft ultra violet light. Then the illusion figures (or any perspective drawings) appear in dramatic depth. This is well worth trying.

There is strictly only one position from which a perspective picture gives correct perspective to the eye. We have some tolerance for error here; but the illusion figures are often drawn with highly exaggerated perspective—strictly requiring an impossibly small viewing distance. For realistic depth to be given by luminous figures, of for example the Ponzo or Muller–Lyer figures, moderate angles should be used, and they should be viewed from roughly the appropriate distance. (The depth of texture-free luminous figures can be shown to an audience, with striking effect.)

We can get over the difficulty that the distortions occur, though the figures appear flat, by supposing that perspective (or other depth cues) sets constancy scaling *directly*.

Although there is nothing implausible in this suggestion, it flouts what has, until recently, been assumed about constancy—that it always follows *seen* depth. This has been held with great authority, as is clear from the following quotation from W. H. Ittleson, who cites support from five other distinguished visual scientists who have worked on this problem. He has this to say: 'Constancy, it is universally agreed, is dependent upon the proper estimation of distance.' We suggest that, although scaling can indeed follow perceived distance (as in the Emmert's law demonstration with after-images), it can also be set directly by depth cues, including perspective. We see this when perceived depth is countermanded by the background texture of pictures. We should then expect distortion illusions to occur when distance is perceived incorrectly—as when there are misleading depth cues. Here we may note that depth cues cannot be appropriate both for the flat surface of the pictures and for the scene or objects they represent. Pictures are inherently paradoxical, suggesting depth yet being flat, so it is hardly surprising that distortions occur.

We suggest, then, there are two kinds of scaling: '*upwards*' from depth cues and '*downwards*' from seen depth.

How can we separate effects of 'upwards' signals from the eyes, from 'downwards' perceptual knowledge or assumptions? We can use phenomena of ambiguity.

Separating bottom-up from top-down by ambiguities

A useful experimental trick is to use perceptual ambiguity for teasing out what is given upwards by cues present in the retinal image, from downwards contributions of knowledge and assumptions. This works because, with ambiguity, perceptions change though the stimulus input remains constant—so top-down effects can be revealed.

Significant phenomena are associated with a flipping Necker cube drawing (Figure 10.6). Much better, is a truly three-dimensional wire cube. (These phenomena are seen most clearly when the cube is coated in luminous paint, to glow in the dark, but this is not essential.)

The skeleton cube will switch in depth, as for the flat figure; but also it will suddenly stand up on one corner, and will rotate with every movement of the observer (at twice the speed), and change shape. These phenomena are worth considerable look-time and think-time, as they can tell us a great deal about how vision is organized: bottom-up and top-down, and as we will suggest, 'side-ways'.

When the wire cube has flipped in depth it moves *with* instead of, as normally, *against* every movement of the observer. For as back and front are depth-reversed, motion parallax is attributed wrongly to near and far (see Chapter 6, page 98). What interests us here, though, is the change of shape. Note that when the cube is seen normally, not reversed, its back and front faces appear equal and all the angles right angles. This is quite surprising, for with an isolated skeleton cube there are no depth cues to set size constancy. It shows that here constancy scaling is operating with no bottom-up depth cues. When it flips in depth, the now apparently back face looks too large, and the angles are no longer right angles. It no longer looks like a cube, but a truncated pyramid, with the further face larger than the nearer. This effect is important. It shows that constancy scaling can be set *downwards*—from depth as seen—rather than signalled (Figure 10.20, Table 10.1).

10.20 Perception of wire cubes

Object perception, one eye

The wire cube looks like a cube. Although the back face gives a smaller image at the eye, it does not *look* smaller. All the angles appear as right angles (though in the projection at the eye they are not—for the projection at the eye is a *perspective projection*). When the cube reverses (like the Necker cube) in depth, then *it no longer looks like a cube*. The apparent back looks too large and the front too small. It looks like a topless (truncated) pyramid. This change of shape upon reversal is very dramatic, and happens every time, with all observers.

What the viewer sees:

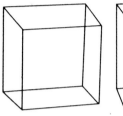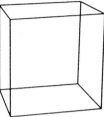

(a)

Picture perception, perspective projection

The cube seen projected as a perspective projection on a screen looks distorted. One face is seen as smaller than its opposite face. The smaller face is seen as the back of the cube, but at the same time it is seen as lying on the screen, at the same distance as the larger (front) face. So depth is paradoxical; as it is a cube and a flat line figure. When it reverses it does not change shape, as does the wire cube viewed directly.

What the viewer sees:

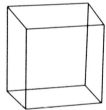

(b)

Zero-perspective object perception

Here we replace the wire cube with a topless (truncated) pyramid. It is viewed from the smaller end, at such a distance that the *retinal image* of the nearer face is the same size as the further face. This is done by making the front face exactly overlap the back face. The object is then rotated slightly to prevent the back being hidden by the front. It is viewed with one eye. The further face looks larger than the nearer face. When it reverses, the *apparently* further face still looks larger than the *apparently* nearer face. So it looks like a topless pyramid. The further face always looks the larger, whichever this may be, when it reverses in apparent depth.

What the viewer sees:

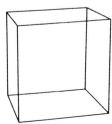

(c)

Picture perception
This is a Necker cube drawing. No face is indicated by perspective as the back or the front. They look the same size, and there is no size change upon reversal. (Because of the texture of the "paper depth".) Depth is paradoxical.
What the viewer sees:

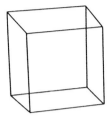

(d)

Stereo-perspective object perception
Here we look at the wire cube with *both* eyes. It looks like a cube; all sides of equal length and all angles right angles, though as for the single-eyed view the image of the back face is physically smaller than the image of the front. It seldom reverses in depth, though it can reverse. When reversed, it is curiously puzzling, not looking quite real. We will have more to say about this effect later. When it does reverse in depth it looks distorted, no longer like a cube. As in the one eye's view, the apparent back looks too large.
What the viewer sees:

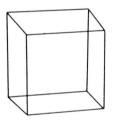

(e)

Stereo-perspective picture perception
We see the picture as three-dimensional, looking incredibly real. It may, indeed be impossible to tell it from the same object (the wire cube) seen directly. It looks like a true cube with no distortion. When reversed (which rarely happens) the stereo picture cube looks distorted, just as for direct viewing of the reversed cube object. If the eyes are switched over: the stereo picture cube looks larger when behind the screen—that is, when apparently more distant from the observer, as in Emmert's Law.
What the viewer sees:

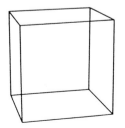

(f)

10.20 (*continued*)

Stereo-zero-perspective object perception
Here we look again at our topless pyramid, but with both eyes. It looks, correctly, like a topless pyramid. The nearer face looks smaller than the front face, as it physically is, though both faces give the same-sized images to the eyes. When reversed (rare) the apparent back looks too large.
What the viewer sees:

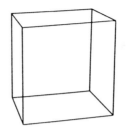

(g)

Stereo-zero-perspective picture perception
The back and front are physically the same size on the screen, but they do not look the same size: the apparent back looks larger. When reversed (rare) the apparent back also looks larger; but when this reversal takes place it has a curious unreal appearance.
What the viewer sees:

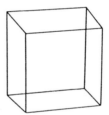

(h)

We have already seen that scaling *can* be set bottom-up from cues to depth, such as converging perspective lines (Figures 10.13). So scaling can be set upwards from retinal signals, or downwards from knowledge or assumptions. These must be very different processes. They can be isolated by making use of phenomena of ambiguity.

Sometimes both upwards and downwards scaling can be involved. An interesting example is the moon illusion—the moon looking larger than usual when it is low on the horizon. This is so though it always subtends the same angle (0.5°) to the eye of an earth-bound observer.

The moon illusion

The moon also looks larger when high in the sky but seen above mountains. It is not the height, but rather the surroundings that affect its

Table 10.1 Conditions giving distortion (yes) and no distortion (no)

When seen correctly in depth

	Mono-perspective	Mono-zero-perspective	Stereo-perspective	Stereo-zero-perspective
Object	no	yes	no	no
Picture	no? (paradoxical)	yes? (paradoxical)	no	yes

When perceptually depth-reversed

	Mono-perspective	Mono-zero-perspective	Stereo-perspective	Stereo-zero-perspective
Object	yes	yes	yes (rare)	yes (rare)
Picture	yes? (paradoxical)	yes? (paradoxical)	no (rare)	yes (rare)

apparent size. The Greek astronomer Ptolomy thought it looks larger because it appears more distant, (what we now call Emmert's law, page 225) but it looks *nearer* and larger. This is unlike Emmert's law.

We may liken it to distortions such as in the Ponzo illusion, where perspective, or other depth cues, scale objects as *larger* when greater distance is signalled, though they do not appear more distant, as the figure lies on a textured background. The moon illusion is not, however, quite the same. For here there is no textured background from the sky to prevent change of apparent distance. What seems to happen is quite complicated: the surrounding depth cues of perspective and so on scale up the size of the moon (bottom-up), so it looks larger. As the sky has no texture—because the moon looks larger, it appears nearer. (This can be demonstrated with luminous disks or balloons in a dark room.) The moon illusion is initiated bottom-up, as for the Ponzo illusion. But this raises a curious question. Why does the moon always appear the same size when high in the sky—and why this particular size? It could be *any* size and distance, from this retinal image free of context. It seems that the visual system has *default* assumptions, which are accepted in the absence of scaling information. (This is like a word processor giving a default line length, in the absence of programmed instructions.) We know very little about this for vision. It could be an important principle.

Distortion by absence of perspective

It is important to note that distortion can occur in the *absence* of perspective—when depth is indicated by other cues, or assumed from knowledge of the object in the picture. In engineering drawings this can be annoying (when perspective is deliberately avoided to show dimensions in isometric projections) for they look weirdly distorted. The effect is seen in this simple perspective-free drawing of a table (Figure 10.21).

We suggest, then, that there are two kinds of scaling: upwards from depth *cues* and downwards from the prevailing *hypothesis* of depth.

Measuring distortion and depth

Visual distortions can be measured in various ways including: matching the distorted perception with (an undistorted) ruler (Figure 10.22); nulling, with an equal and opposite distortion; by lengthening or bending or shortening the figure until it appears undistorted.

Measuring visual depth may seem to be impossible. But consider Figure 10.23.

If the figure has perspective, or other depth cues, it is found that the observer places the marker light not at the true distance, but at the apparent distance of this part of the picture, as he or she sees it. For people with normal stereoscopic depth perception this is quite an easy task, giving consistent measures of apparent distances. So, the visual worlds of observers can be plotted in three dimensions in physical space.

This technique shows that the illusion figures are seen in depth according to their perspective features—when depth is not countermanded by texture of the picture plane. It is found that increased *distance* of features correlates highly with illusory *expansion* in the

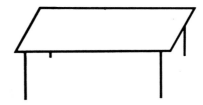

10.21 A top-down distortion. When this is seen as a table lying obliquely in space, the further edge looks too long, and the sides do not look parallel, though this is how they are drawn. This is due to size scaling working downwards from assumed depth.

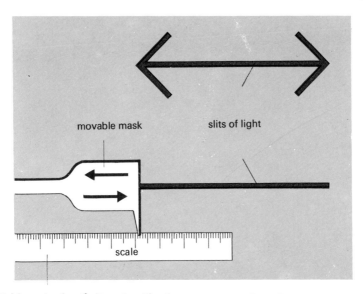

10.22 Measuring length distortion. The observer sees a single Muller–Lyer figure and an adjustable comparison line, set to appear the same length as the distorted line. This gives a direct measure of the distortion. (Back view of the apparatus.)

perspective illusion figures. So we have objectively related these distortions to perspective depth.

The finding that apparent depth is closely related to these distortions is suggestive; but it is hardly proof that perspective depth features produce the illusory expansion. Can we think of a more direct test to distinguish between distortions due to errors of the *signals* of the visual channel and errors due to *information* (such as perspective) applied inappropriately? We may have this in an experiment carried out by the author with John Harris, in which the distortions were found to vanish when the viewing conditions for the flat figures match the normal conditions for seeing equivalent objects in three-dimensional space.

Illusion-destruction

If the distortions of these figures are due to inappropriate constancy scaling—then the distortions should no longer occur when all the features setting the scaling are *appropriate*. To make them appropriate, we ensure that the perspective of the figures is exactly the same at the eyes as for viewing an object (such as corners or receding parallel

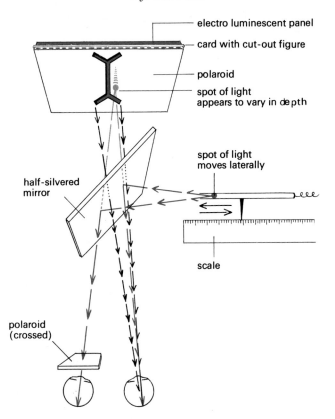

10.23 Measuring apparent depth—'Pandora's box'. The flat figure is back-illuminated, to avoid texture and so avoid paradoxical depth. The light from the figure is cross-polarized to one eye, so it is only seen, say, by the right eye. A small adjustable marker light is optically introduced into the figure, by reflection from a part-reflecting glass or perspex sheet. The movable reference light is seen with *both* eyes. The observer sets it to the apparent distance of any feature of the figure, and its distance is recorded. Thus binocular vision is used to measure monocular depth, for any depth cues. (This can be used with a computer screen).

lines) directly. We further ensure that it is seen in its correct depth—as though it were an object, although it is in fact a flat picture. Consider the Muller–Lyer figure (Figure 10.16): this is a flat perspective projection of corners. We make three-dimensional wire model corners, and project them in accurate perspective by casting their shadows on a screen with a small bright light source (Figure 10.24). When this is viewed with the eyes at the same distance as the light source, the perspective of the retinal image is exactly correct—it is the same as when viewing the object directly. But it will still appear flat. To give the correct three-dimensional perception we add a second shadow-casting light, separated horizontally from the first, by the separation of the observer's eyes. Finally, it is so arranged that one eye sees one of the perspective shadow projections, and the other eye sees only the other projection, by cross polarization. The projections are combined by the brain—to give a three-dimensional perception exactly as though it were a real corner, though it is flat on the screen.

In this situation, the distortion is entirely absent. It is absent although the visual system is signalling the same angles as in the usual illusion figure, appearing flat and distorted. So how can the illusion be due to distortions of (bottom-up) signals of the visual channel? This seems to be clear evidence that the distortions are produced by depth cues when they are inappropriate to the depth of the object, and when distances are seen incorrectly. The first kind of error is incorrect scaling 'upwards' from depth cues; the second is incorrect scaling 'downwards' from perceived depth. This is shown in Figures 10.25 and 26.

It turns out that all the perspective illusions are destroyed when presented in appropriate depth. The simplest and easiest example is the Poggendorff illusion.

Distortions similar to those of vision are also found for touch. This presents something of a problem for the theory we have been advocating. John Frisby has provided evidence that people with vivid visual imagery tend to have greater than usual touch illusions. Possibly touch information is interpreted according to visual processing and visual 'models' of the world. Somehow the senses interact and combine to give generally consistent perceptions of objects.

Paradoxes

Perhaps the first impossible figure was devised by the Swedish artist Oscar Reutersvärd in 1934. The best known visual paradox is the impossible triangle, designed by the English geneticist Lionel Penrose and his nephew the cosmologist Roger Penrose in 1958. A three-dimensional model which appears impossible was shown in my Royal Institution Christmas Lectures in 1967 (published in *The intelligent eye* in 1970). This shows that a normal, unpainted object can appear impossible (Figure 10.27). So it does not depend on artificial combinations of visual cues, though the viewing angle is critical.

From another viewing position it does not appear impossible (Figure 10.28).

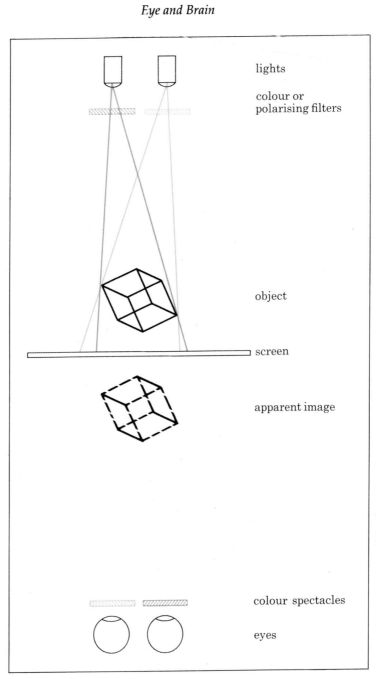

lights

colour or
polarising filters

object

screen

apparent image

colour spectacles

eyes

10.24 3-D shadow projection. A pair of small bright lights (low-voltage halide lamps, in a light-tight box) are horizontally separated by the distance between the eyes. They are cross polarized to the eyes or red–green filters and glasses are used. The illusion figures are made as three-dimensional wire models. With one light their shadow is just the same as the illusion figures—appearing distorted as usual. When the second light is switch on the shadow is immediately seen in the three dimensions of the wire model. The distortion disappears. (This makes a super lecture demonstration. A silver screen should be used to prevent depolarization; or red–green filters and glasses work with any screen.)

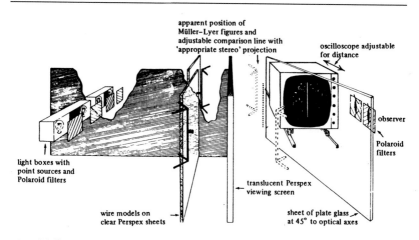

10.25 Muller–Lyer destroyed—apparatus. Horizontally separated small light sources project three-dimensional shadows of wire corners. The separation, giving disparity, and the viewing distance (correct at 40 cm) can be varied.

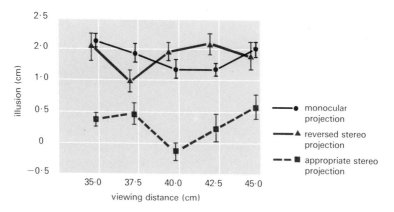

10.26 Muller–Lyer destroyed—results. The Muller–Lyer distortion entirely disappears, when both the perspective and the depth are appropriate. (This is when the viewing distance is 40 cm—the same as the distance of the wire model corners from the point sources—and they are separated by the separation of the eyes (Gregory and Harris, 1975).)

10.27 Impossible object, made of wood. It appears impossible from a critical position, from which the photograph is taken.

The problem arises because of a quite powerful visual rule: that when features or objects are *touching* they tend to be seen at the same *distance*. This, though they could well be very different distances. (Binary stars appearing close together may be separated in distance by millions of light years.) It is striking that when any corner of the impossible triangle is hidden, it appears perfectly possible (Figure 10.29).

This illustrates how following a rule, from a false assumption, can generate a paradox. This applies to conceptual and perceptual hypotheses.

10.28 Impossible object—seen as possible. The same wooden model, photographed from different positions. It no longer looks impossible. But although this gives the answer away, Figure 10.29 still cannot be *seen* as possible!

Fictions

Vision can take off as fiction—illusory contours and surfaces being seen by all observers in normal viewing conditions. The most famous is the Illusory Triangle of the Italian artist–psychologist Gaetano Kanizsa (1974) (Figure 10.30).

An earlier though less impressive 'fictional' figure was shown by F. Schumann in 1904 (Figure 10.31).

10.29 The impossible triangle with a hidden corner appears perfectly possible, whichever corner is hidden.

Here another example, using more top-down knowledge, is shown in Figure 10.32. Although shadows are joined to, and so optically are parts of objects, they are perceptually quite distinct and are very seldom confused with objects. Shadows are so powerful as visual cues that they can evoke perception of objects when there are no objects present. This is clear in the typeface of Figure 10.32. Here we see letters as large as life; but in fact, only shadows of imaginary letters are present.

Using a simple figure similar to Kanizsai's triangle (Figure 10.33), John Harris and the author found that the fictional surfaces are only seen when they are perceptually *in front* of the inducing figure.

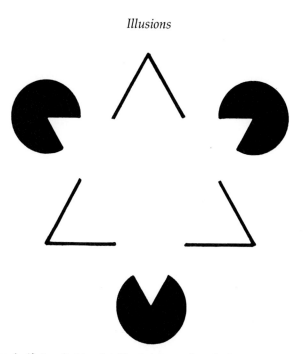

10.30 Kanizsa's 'fictional' triangle. The brighter—than background—triangle is not physically present.

10.31 Schumann's visual 'fiction' was shown in 1904—but, curiously, was ignored before Kanizsa's more dramatic examples.

SHADOW

10.32 Letters? These are merely shadows, but we see the objects which would cast these shapes as shadows. The brain invents missing objects on the witness of usually reliable cues.

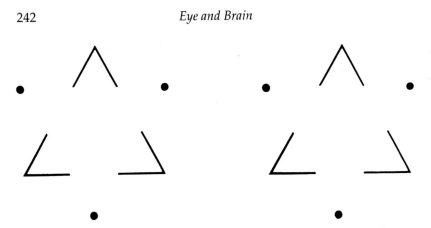

10.33 Illusory surface destruction—by stereo depth. When seen in a stereoscope, the dots go behind the lines. Then the illusory surface and contours disappear. (They must lie in front to 'explain' the gaps).

When forced behind, by stereopsis, they disappear. This is evidence that they are visual *postulates* accounting for, and suggested by, the gaps. So occlusion is both an important cue to depth, and is postulated when a nearer occluding object is highly likely even though not present.

We see that surprising *absence* of stimulation can serve as *data* for perception. Similarly, the illusory letters seen in Figure 10.32 are postulated from the shadows. Even simple perceptions, such as contours and surfaces, can have 'cognitive' origins. These are probably not at a very high level of processing. Specific *object* knowledge does not seem able to create missing parts—such as missing noses—so seems to operate by fairly simple low-level rules.

This account is not universally accepted. Some authorities suppose that these are contrast effects, or are given by neural interactions such as lateral inhibition. The issue is whether these illusory contours are produced directly by *stimuli*; or far less directly by surprising absence of stimuli, which conveys information.

It is suggestive that fictional contours presented in a stereoscope and combined by the two eyes, can give three-dimensional depth as for true contours or lines. Experiments of this kind are important, to establish what is 'physiological' and what is 'cognitive'. This distinction, though hard to make, runs right through psychology and also through the practice of medicine. Sorting these issues out for clear-cut phenomena of perception may be helpful for thinking about difficult clinical judgements.

This has been a complicated journey. Not everyone will accept the route taken, or the conclusions of the arrival. This is the controversial way science is—and to keep alive and exciting, should be. We will try to bring these difficult issues together in Chapter 11, with speculations at the edge of our understanding of eye and brain, including a tentative classification of the wonderful phenomenal phenomena of illusion.

11

Speculations

We have hardly said anything about how emotions may affect seeing, except for the empathy theory of distortions, which we dismissed in favour of very different ideas (page 208). Perhaps, now, we should listen to Lady Macbeth:

> Is this a dagger which I see before me,
> The handle toward my hand? Come, let me clutch thee:
> I have thee not, and yet I see thee still.
> Art thou not, fatal vision, sensible
> To feeling as to sight? or art thou but
> A dagger of the mind, a false creation.
> Proceeding from the heat-oppressed brain?
> Mine eyes are made the fools o' the other senses,
> Or else worth all the rest: I see thee still
> And on the blade and dudgeon gouts of blood,
> Which was not so before. There's no such thing:
> It is the bloody business which informs
> Thus to mine eyes.

Emotions can affect perception, though not systematically (as in the illusions discussed in the previous chapter). Conversely, perceptions can affect emotion—hence powers of art. Some things look beautiful, others ugly. But after all the theories that have been put forward we have little idea why this should be so. The answer must lie far back in the history of human experience—no doubt from pre-human dramas of successes and failures of survival. For presumably we find flowers and sunny days beautiful because, through millions of years, they enhanced living and increased chances of creating new life. So we resonate with flower-seeking insects, but hardly with creatures of the dark. Our appreciation of symbols has immeasurably extended ancient responses; so surrogate pictures and words and music evoke ancient emotions related to surviving and reproducing ourselves.

It has been suggested that emotions are sensations of *bodily adaptations* to situatioins of danger, love, or whatever. This is known as the James–Lange theory, from the American psychologist William James (1842–1910) and Swedish physiologist C. G. Lange (1834–1900). They suggested that emotions do not cause bodily changes, but rather the other way round. For example a flood of adrenaline for the 'flight or fight' reaction to a dangerous situation causes the sensations of fear. This means that feelings of emotions are perceptions of reactions of our bodies, to situations they responded to aeons ago. Although this may not explain all the subtleties of emotions, it does seem to be their biological basis.

In this account, emotions are intimate perceptions of bodily changes; but their significance is very much in terms of social situations and how they are perceived. Charles Darwin suggested that the red cheeks and neck of blushing are a public signal that the individual has transgressed a social rule and so is not to be trusted. The red flush gives the game away, while the accompanying sense of confusion makes it hard to concoct an excuse. Blushing is unique to humans and does not occur in childhood before social mores for 'good' and 'bad' behaviour are appreciated.

Whatever their ancient origins, emotions are deeply associated with meaning. A recurring theme of this book is the importance of meaning for perception—and also of sophisticated perception for perceiving meanings. This develops from relatively meaningless *patterns* to seeing potentialities of *objects*. Object-meaning transfers to pictures, though objects seen in pictures are powerless to hurt or to reward, and we are but passive spectators of pictured worlds. Objects are defined initially by what they can do to us and we to them. Almost whatever a table looks like it is an object for putting things on, and a chair is for sitting on. Certain expectations must be fulfilled. If a book were placed on a supposed table which melted away, we would say that it was not after all a table, or any object. It might be written off as a dream or hallucination. Here mirror images are interesting, for though objects are seen, they are dislocated in space and divorced from behaviour. All *pictures* of objects are transposed in space and time.

Although the sensory dimensions of sight, touch, smell, and so on, are very different, we have no hesitation in accepting that they all belong to the same world of objects. And clearly our knowledge of objects is not limited to sensory experience. We know about magnetism though we cannot sense it, and about atoms though they are invisible. An essential power of science is to extend knowledge far beyond sensory perception—often to change how we see.

It is not hard to guess why the Intelligent Eye creates hypotheses beyond visual information; as behaviour can be directed not only to what is sensed but also to what is hidden, and to what may happen in the immediate future. The richness of perceptual hypotheses (perceptions) confers immense survival benefits, as well as making the world and illusions perceptually and conceptually interesting.

If the brain were unable to fill in gaps, and to bet on limited evidence, behaviour would come to a halt in the absence of directly relevant data from the senses. The cost of going beyond the evidence—to see and live by predictive hypotheses—is a rich variety of illusions generated by intelligence. Intelligence is dangerous, but stupidity is worse! Lack of intelligence is worse because it limits possibilities. We might say that perceptual intelligence modifies the computer adage 'Garbage in—garbage out', to 'Garbage in—*sense* out'. But evidently this is far from infallible, for many illusions and errors are generated by seeking sense from sensation.

Let's try to bring our too-complicated discussions of phenomenal phenomena of illusions to a focus—by attempting to classify them. Why should we go to this further trouble? Classifying is extremely important for any science. Mendeleyev's periodic table of the elements made sense of chemistry, its gaps suggesting where to look for new advances. Classifying species in biology, and kinds of stars in astronomy, led to understanding their evolution. Classifications can separate useful from misleading analogies. They structure present knowledge and they guide new observation and experiment.

Even more basic: at least implicit classes are needed for building up generalized knowledge from instances by *induction*. Explicitly defined classes are also essential for *deduction*—for inferring particular conclusions from generalizations, with formal rules.

The simplest animal learning is inductive, including Pavlovian conditioning—such as relating a buzzer to food. No doubt this is why much of learning is slow: relevant associations have to be discovered through many instances. Rules of deduction are relatively recent: formulated by Aristotle for his syllogisms.

Possibly this attempt to classify phenomenal phenomena of vision will help to structure the unnatural science of illusion. Its gaps and limitations may suggest observations and experiments, for further understanding of how we see. So, we will end by:

Putting illusions in their place

Classifications change with shifts of theoretical understanding or opinion. An almost too obvious biological example of this is the whale. Looking like a huge fish, it came to be classified as a mammal, as it is air breathing, bears its young, and so on. Initial appearances give way to needs of classification. More detailed classes are developed when more discriminations are needed, generating technical terms which, though useful for experts can be barriers to general communication. This is especially so when the criteria for belonging to classes are not simple appearances. Here, the exercise of trying to classify should make our understanding more explicit, and it may suggest 'litmus tests' for deciding where to put phenomena of perception in their place.

The appearances fall neatly into four classes, which we have called: *ambiguities, distortions, paradoxes, fictions.* It may be noted that these are also errors of language. This may be no accident. Perhaps human language developed from pre-human perceptual classifications of objects and actions. This notion might explain why the natural languages have similar basic structures. This is stressed by experts such as Noam Chomsky and Steven Pinker, though such innateness is perhaps controversial.

We start with the physical world: the source of signals to the senses. The first source of visual illusions lies in disturbances of light before the eye is reached. These we may call *physical* illusions. Though whether there is an illusion depends on whether or not the perceiver corrects for the optical disturbance.

Sensory signals are carried by a great variety of parallel neural channels—for contours, movement, colours, stereo depth, and so on—and are processed by specialized brain modules (Chapter 4). Upsets of the physiological channels or the brain processing modules may generate *physiological* illusions.

The signals are 'read' in two ways, to give useful data: through general *rules*, and by *knowledge* of objects. Using rules and knowledge can be called 'cognitive.' When rules or knowledge are misapplied, illusions occur through no 'fault' of the physiology. So although physiology is always involved, it is not always responsible for errors or illusions. It is what the physiology is doing—and whether this is appropriate—that matters.

We may call the signals from the senses 'bottom-up', and the application of knowledge (which may or may not be appropriate) 'top-down'. Here I shall add 'sideways' for rules. (This by rough analogy with putting floppy disk programs 'sideways' into a computer; though we are not assuming the brain to be a digital computer. It seems, rather, to be analogue modules of neural nets.)

We have hinted that perhaps the essential structure of languages derived from ancient, pre-human, perceptual classifications of objects and actions. For it is striking that the obvious names for kinds of illusions are the same as for errors of language: ambiguities, distortions, paradoxes, fictions (Table 11.1).

Table 11.1 Comparison of language errors and visual illusions

	Language errors	Visual illusions
Ambiguity	*People like us*	Duck–rabbit (Fig. 10.8)
Distortion	*He's miles taller than John*	Muller–Lyer (Fig. 10.16)
Paradox	*She's a dark haired blonde*	Impossible object (Fig. 10.27)
Fiction	*They live in a mirror*	Kanizsa triangle (Fig. 10.30)

We have found also four basic causal classes of illusions: *physical, physiological, knowledge, rules*. Let's construct a classification—in terms of appearances and kinds of causes—which might be helpful for understanding, and for suggesting new observations and experiments.

Physical optical illusions have their cause before the eye is reached. They do not tell us anything about physiology or cognition; except, indeed, that they are not countered by cognitive visual processes, or checks, or by understanding of the situation. The lack of checks is presumably because perception has to work extremely fast to be useful in real time. (Knowledge, for example of reflections, does not counter mirror illusions.)

In general, we might say of phenomenal phenomena that *physiological* illusions tell us about *brain*, and *cognitive* illusions about *mind*. (This, if we think of mind as the brain's knowledge and hypothesis-generating rules and intelligence.)

Classes of illusions

Table 11.2 shows a tentative classification in these terms of phenomenal phenomena of vision. It represents tentative decisions for putting

Table 11.2 Classification of phenomenal phenomena

	Ambiguity	Distortion	Paradox	Fiction
Physical input	Mist; confusing shadows	Stick in water (of space); stroboscope (of speed)	Mirrors (seeing oneself double in a wrong place)	Rainbows; moiré patterns
Physiological signals	Retinal rivalry	Café wall; adaptations to length, tilt, curvature	After-effect of motion (moving, yet not changing position or size—when neural channels disagree)	After-images; migraine patterns
Top-down knowledge	Duck–rabbit; vase–faces; Necker cube	Size–weight illusion (small objects feel heavier than larger objects	Pictures (seeing objects with patterns of paint)	Faces in the fire; ink blots
Sideways rules	Figure–ground	Muller–Lyer, Ponzo, Orbison, Poggendorff (perspective setting size scaling inappropriately)	Impossible triangle; Escher's pictures	Kanizsa triangle; filling in of blind spots and scotomas

these phenomena in their place. How can we decide more objectively? We need more explicit 'litmus tests' for distinguishing and classifying phenomenal phenomena. I hope to develop this in a forthcoming, more technical book on illusions.

Design for mind

If we put these various ideas together, we arrive at a scheme something like Figure 11.1. (This is a development of Figure 10.9.)

Perceptual brain systems do not always agree with the thinking cortex. Thus, for the cortex educated by physics, the moon is about 400 000 km distant (nearly a quarter of a million miles), but to the visual brain it appears but a few hundred metres away. Although we know the distance intellectually, the visual brain is never informed, so we continue to see the moon as almost within our grasp. This deep separation between conceptions and perceptions is disturbing, yet no doubt inspires much of science and art.

In spite of many experiments—including Stratton's experience with inverting goggles (page 141), Held and Hein's kittens in baskets (page 144), and adult recovery from infant blindness (pages 151–158)—we still know little of how perceptual learning works; yet relations between perceptual and conceptual learning are obviously important for education. Interacting—playing—with objects develops perceiving and conceiving. So, hands-on science centres may help children and extend perceptual and conceptual learning through adult life. Figure 11.1 shows feedback from handling objects to develop perceptual and conceptual knowledge. Much of our normal hands-on experience from infancy onwards is, however, misleading for understanding physics (tending to make us think in Aristotelian terms) because, for example, toy cars (and real cars) need a continuous force to keep them going. This, because friction contaminates and hides the deep insights of Galileo's physics and Newton's laws of motion. Hands-on science centres should allow basic laws of physics to be experienced as purely as possible, to relate perception to conceptual understanding. Frictionless toys should be designed and given to babies!

The behaviourists of the 1920s tried to deny consciousness. Most of the phenomena discussed in this book are in our awareness—in our consciousness. Although behaviourism is now a dead doctrine, behaviour itself is of course extremely important, including as an output of visual perception. There would be little point in seeing if we could not act on what we see. What is surprising is how much more

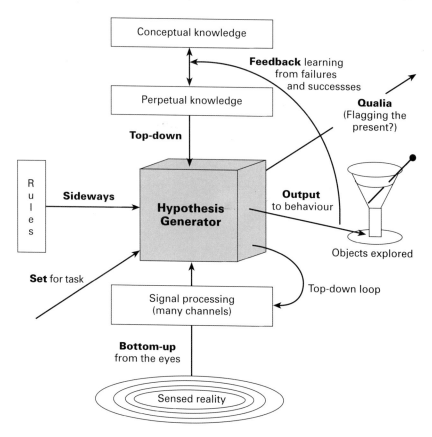

11.1 Speculative mind-design for vision. Bottom-up signals from the eyes, and other senses, are processed physiologically and interpreted or 'read' cognitively by object knowledge (top-down) and by general rules (side-ways). The general rules (such as perspective and Gestalt laws of organization) are syntax; the object knowledge is implicit semantics. Feedback from successes and failures of action may correct and develop knowledge. (Hence the importance of hands-on learning.) It is suggested that real-time sensory signals flag the present—conceivably with *qualia* of sensation.

there is to perception than appears in behaviour, including appreciation of beauty in nature and the arts.

Now there is new interest in processes of vision concerned with action. Mortimer Mishkin among others suggests that there are two cortical pathways for vision, one ventral (occipitotemporal, linking striate cortex to prestriate regions, then to inferotemporal cortex); the other dorsal (linking the prestriate regions to the posterior parietal lobe). Interruption of the ventral ('downwards') path abolishes visual object discrimination; interruption of the dorsal path from the eyes ('bottom-up') disturbs spatial vision, including errors of actions involving positions of objects.

This relates to the distinction between 'seeing what' and 'seeing where'. Brain lesions can affect one and not the other. These brain circuits may also be related to implicit and explicit knowledge (which is hard to define apart from language) and perhaps to what is or not conscious.

There is much of great interest on conscious–visual and behaviour–action pathways in David Milner and Mark Goodales's (1995), and in Marc Jeannerod's (1997) recent books, with promise of exciting research to come on relations between mind and brain. No attempt is made in this book to represent anatomical brain systems in the scheme 'Ins-and-Outs of vision' (Figure 11.1), as this is quite abstract and so does not reflect brain anatomy or physiology. This awaits further research on relating structures to cognitive functions, with brain imaging with PET scans (page 81), and also deeper theoretical insights into how the brain works. For we must know what functions are involved, to recognize or localize them.

Evidence that there are two visual systems comes from the surprising finding that some distortion illusions are greater as *seen* than in *behaviour*. Thus, the two inner circles of the Titchener illusion (Figure 8.10) look different sizes; but reaching and grasping actions are not affected by the visual illusion produced by the different sizes of the outer circles. This separation of seeing from doing has implications for interpreting observations on animals and infants, when language is not available. It makes it even harder to discover what they are seeing from behaviour.

Let's end with a few thoughts on:

What is consciousness?

Conscious experience—sensations of red, cold, pain, and so on— are our most immediate knowledge and yet are utterly mysterious. In religions and most philosophy, consciousness is supposed to be separate from the brain, which is seen as a receiver of external transmissions. The prevailing view in the brain sciences is very different: that structures and functions of the brain generate mind and consciousness. Francis Crick calls this the 'astonishing hypothesis'. Historically astonishing and repugnant, this is now what most of us believe. In the mid-18th century the French doctor–philosopher Julien

Offray de la Metrie was ostracized and forced to leave his home and work in Paris for just this view, expressed in his book *L'homme machine* (1748)—yet now this is the received wisdom, at least of the brain sciences.

A striking difficulty is that sensations seem so different from anything physical. It is hard to imagine how physiological activity of brain cells could produce sensations—now often called *qualia*—such as red or pain. The gap between physiology and sensation seems just too great to bridge by intuitive imagination or by experimental science. But (almost by definition) the most important discoveries of science are counter-intuitive. This why it is unsafe to apply common sense to science, which no doubt can make science frightening.

It is worth pointing out that there are surprising gaps within physics—for example the result of Michael Faraday's experiment of 1831 which produced electricity by moving a magnet through a coil of wire. Electricity looks incredibly different from moving wires and magnets. Here science gradually developed linking concepts, so this gap is almost filled. Even more puzzling is the timelessness of the equations of physics—yet we know we are living in irreversible time.

In spite of centuries of thought and experiment, gap-filling has not so far been achieved for linking physical brains to mental consciousness. We still hardly know which brain functions are significant. Perhaps even worse, we don't know how to recognize consciousness in other animals (especially animals very different in looks and behaviour from ourselves), or in a robot. The great difficulty here is that we do not know what—if anything—consciousness does behaviourally.

Here are three hard questions of consciousness:

1. How can physical structure and processes of the brain produce awareness—sensations or *qualia*?

2. Do qualia have causal effects on brain states or behaviour? (In which case physics and physiology could not be adequate for explaining vision.)

3. If qualia have effects—what do they do?

A dilemma is: if consciousness *is* causal, we need to incorporate it into accounts of behaviour; if it is *not* causal, and so no use—why did it evolve through natural selection? It is very hard to see how sensations—qualia—can fit into science if they have no causal consequences, and so cannot be recognized 'from the outside'. An interesting, controversial move is Roger Penrose's suggestion to extend quantum physics' somewhat mind-like characteristics to reduce the gap. If consciousness is at the level of fundamental particles and processes of physics: why isn't all matter—tables and chairs—conscious? Penrose pin-points certain very small biological structures, microtubules, as specially involved in consciousness; but then we need to ask why (assuming this is so) amoeba are not conscious. In short— what is so special about the brain? It is interesting that our consciousness can be cut off with anaesthetics. If we knew more of how they worked anaesthetics might provide clues, perhaps combined with brain scan experiments.

Whether new techniques for recording brain activity (PET and NMR scans, page 81) will provide basic answers is an exciting possibility. At least they may show where consciousness-generating activity is located in the brain, and they can show which regions are active, not only from sensory stimulation, but according to cognitive processing. There are now criteria for relating physiological brain activity to what it is doing cognitively.

Consciousness is certainly hard to think about. But for the physicist, matter and force and time are perhaps as puzzling, if in different ways. Science is good at describing relations—powerless to handle uniqueness. So perhaps the question 'What is *matter*?' is as hard to answer as 'What is *consciousness*?' because both lack analogies to anything else. It is comforting that consciousness is not the only mind-boggling problem!

May we hazard a guess for what qualia might do? A central notion of this book is that perceptions are but tenuously related to reality; being predictive hypotheses, having major contributions from top-down knowledge and sideways rules derived from the past. Now we may see a problem.

Hypotheses do not especially refer to the *present*—yet the *present moment* is crucial for survival. There is no such problem for simple reactions to stimuli—reflexes and tropisms; for stimuli come bottom-up from the present. With development of perceptual hypotheses late in evolution, sensory inputs become relatively less important—as they are greatly enriched by stored knowledge and general rules. Could it be that stimuli retain their original role of signalling the present—but are now *flagging* the present moment with *qualia*?

It is suggestive that vision is qualia-rich compared with immediate memory and imagination. Try looking at something, then shut your eyes and imagine it—the difference is surely amazing. The qualia fade in an instant. Memories are far less rich in qualia than perceptions. Could this be because memories lack sensory signals—necessary for signalling the present moment?

An exception is memories of emotions. One may blush with shame, when re-living an embarrassing moment in memory. If the James–Lange theory of emotion is at least in part correct—that emotions are sensations of bodily reactions (page 245)—it could be that remembered emotions trigger qualia from the sensed bodily changes. Qualia of vivid dreams, and schizophrenic hallucinations, might be where this supposed qualia-flagging system for identifying the present breaks down. Conceivably this has clinical implications. This qualia idea is speculative; but as the tortoise said, 'I can't take a step forward without sticking my neck out.'

A heavily top-down robot would need to use its real-time inputs to signal the present moment; but would it need qualia? Should we ever come to confer our processes of perception to machines, perhaps problems of consciousness will be solved. Or, perhaps we need to answer qualia questions to be able to build conscious machines. Meanwhile, as for any science, we may continue to study and try to explain phenomena without 'ultimate' understanding. This is how science works, as any explanation has limits.

The physical sciences take immense trouble to avoid errors. Here we seek out and study errors for understanding how we see and to suggest something of how the brain works. The weird and wonderful errors of illusions are not trivial. They are truly phenomenal phenomena, central to art and a major reason for the experimental methods of science.

Bibliography and notes

General

Texts

Barlow, H. B. and Mollon, J. (1982). *The senses*. Oxford University Press. [Technical, authoritative account, aimed at medical students]

Bruce, V., Green, P. R., and Georgeson, M. (1996). *Visual perception: physiology, psychology, and ecology*, 3rd edn. LEA, London.

Frisby, J. P. (1979). *Seeing: illusion, brain, and mind*. Oxford University Press. [Excellent on receptive fields and physiology]

Gross, R. D. (1992). *Psychology: the science of mind and behaviour* (2nd edn). Hodder and Stoughton, London. [Massive, well written general text]

Howard, I. and Templeton, W. (1966). *Human spatial orientation*. Wiley, New York. [A classic, on these generally under-presented topics]

Hubel, D. H. (1988). *Eye, brain, and vision* (Scientific American Library). W. H. Freeman, New York. [Clearly written with excellent illustrations, by the Nobel Laureate who, with Torstin Wiesel, discovered cortical 'feature detectors']

Kaufman, L. (1974). *Sight and mind*. Oxford University Press.

Neisser, U. (1967). *Cognitive psychology*. Appleton-Century-Crofts, New York. [Pioneer book on the subject of its title]

Pirenne, P. H. (1948). *Vision and the eye*. Chapman and Hall–Anglobooks, London. [Extremely clear technical account of photons and vision]

Rock, I. (1975). *An introduction to perception*. Collier MacMillan, New York.

Rock, I. (1983). *The logic of perception*. MIT Press, Cambridge Mass.

Rock, I. (1984). *Perception* (Scientific American Library). W. H. Freeman, New York. [The late Irvin Rock treats vision as 'problem solving': the essentially Helmholtzian philosophy of the present author. He was a most ingenious experimenter]

Sekuler, R. and Blake, R. (1994). *Perception* (3rd edn). McGraw Hill, New York. [Technical physics-based treatment]

Woodworth, R. S. and Schlosberg, H. (1954). *Experimental psychology*. Holt, Rinehart and Winston, New York. [A classic standard, though now dated, general text]

Historical

Al-Haytham, Ibn (Alhazen) (*c.* 1040). *Optics* (trans. A. I. Sabra, 1989. *The Optics of Ibn Al-Haytham*). The Warburgh Institute, University of London. [The first account of linear propagation of light and first systematic discussion of illusions as phenomena]

Boring, E. G. (1942). *Sensation and perception in the history of experimental psychology*. Appleton-Century-Crofts, New York. [A classic history]

Boring, E. G. (1950). *A history of experimental psychology*. Appleton-Century-Crofts. New York. [Also a classic.]

Galton, F. (1883). *Inquiries into human faculty*. Macmillan, London. [Galton pioneered statistical measures of human anatomy and performance, and discussed mental imagery]

Gregory, R. L. (1981). *Mind in science*. Weidenfeld and Nicolson, London. [Compares the history of psychology and physics back to the Greeks. Technology seen as important for suggesting and testing ideas]

Helmholtz, H. von. (1856–67). *Handbuch der physiologischen Optic*. L. Voss, Hamburg. 3rd edn (1909–1911) *Helmholtz's treatise on physiological optics* (trans. ed J. P. C. Southall). Optical Society of America, New York. Translation reprinted (1962). Dover, New York. [The founding—and still highly valuable—study of vision]

James, W. (1890). *Principles of psychology*. Holt, New York. [Classical account of psychology. Remains important]

Lucretius (*c*. 80 BC). *De rerum natura*, (trans C. H. Sisson (1976). Carcanet Press, Manchester. [Celebrated poem outlining Greek and Roman science, especially atomism]

Mach, E. (1897). *Beiträge zür Analyse der Empfindung*. (*Analysis of sensation*. Dover, New York. [Austrian physicist who made insightful comments on several visual phenomena]

Necker, L. A. (1832). Observations on some remarkable phenomena seen in Switzerland, and an optical phenomenon which occurs on viewing a crystal or geometrical solid. *Phil. Mag.*, (**3 ser.**), 329–37. [First account of 'flipping' ambiguities; though these were probably known far earlier, as in Roman tile designs]

Rubin, E. (1915). *Synsopplevede Figure*. (trans. S. M. Sherman, 1985). In *Progress in psychobiology and physiological psychology*, Vol. II (ed. J. M. Sprague and A. N. Epstein), pp. 233–322, Academic Press, New York. [First account of face–vase ambiguity]

Schumann, F. (1900). Contributions to the analysis of visual perception—First paper: Some observations on the combination of visual impressions into units. In *The perception of illusory contours* (1987), (ed. S. Petry and G. E. Meyer), pp. 21–34. Springer, New York. [First example of illusory contours, though may have been known to artists much earlier]

Warren, R. M. and Warren, R. P. (1968). *Helmholtz on perception*. John Wiley, New York. [Useful short account of Hemholtz's immense contribution]

Watson, J. B. (1930). *Behaviourism*. Norton, New York. [Classical account of attempt to deny consciousness, perhaps to make psychology more 'scientific']

Philosophical

Berkeley, G. (1704). *A new theory of vision* (Everyman's Library). Dent, London. [These essays—including 'The three dialogues between Hylas and Philonous, and 'A treatise concerning human knowledge'—written while young, with verve and style, raise questions that are still discussed]

Boden, M. A. (1972). *Purposive explanation in psychology*. Harvard University Press.

Boden, M. A. (1988). *Computer models of mind*. Cambridge University Press. [Philosophy written by a knowledgeable psychologist: raises central issues of mind and computers]

Churchland, P. (1988). *Matter and consciousness*. MIT Press, Cambridge, Mass. [Exceptionally clear exposition of difficult ideas]

Churchland, P. S. (1989). *Neurophilosophy: towards a unified science of the mind–brain*. MIT Press, Cambridge, Mass. [A technical 'reductionist' account of brain-based mind]

Dennett, D. (1991). *Consciousness explained*. Little, Brown, New York.

Descartes, R. (1637). *Discourse on method* (Trans. P. J. Olscamp (1961)). Bobbs-Merrill, Indianapolis. [Classic statement by the founder of 'dualism'—with early scientific insights on optics and eyes]

Kuhn, T. (1962). *The structure of scientific revolutions*. Chicago University Press. [Probably the most important modern account of philosophy of science. Emphasizes importance of basic assumptions—'paradigms'—and what happens to perceptions of science when they change]

Ryle, G. (1949). *The concept of mind*. Hutchinson, London. [Classic Oxford philosophy, critical analysis of notions of mind. Tends towards behaviourism]

Wittgenstein, L. (1958). *Philosophical investigations* (2nd edn). Blackwell, Oxford. [Major statement of genius. Specifically insightful on visual ambiguity]

1. Visions of vision

Barlow, H., Blakemore, C., and Weston-Smith, M. (eds) (1990). *Images and understanding*. Cambridge University Press.

Blakemore, C. (1988). *The mind machine*. BBC Publications, London.

Blakemore, C. and Greenfield, S. (1987). *Mindwaves*. Blackwell, Oxford.

Bruner, J. S. (1974). *Beyond the information given*. Allen and Unwin, New York. [Emphasizes use of knowledge for perception]

Craik, K. J. W. (1943). *The nature of explanation*. (Cambridge University Press. [Pioneer account of perception as representing]

Ellis, W. H. (1938). *Source book of Gestalt psychology*. Routledge and Kegan Paul, London.

Fodor, J. A. (1983). *The modularity of mind*. MIT/Bradford Books, Cambridge, Mass. [Re-assesses phrenological concepts]

Gibson, J. J. (1950). *The perception of the visual world*. Houghton Mifflin, Boston.

Gibson, J. J. (1966). *The senses considered as perceptual systems.* Houghton Mifflin, Boston. [Classical statements of Gibson's 'direct' account of vision]

Gordon, I. E. (1997). *Theories of visual perception.* (2nd edn). Wiley, New York.

Gregory, R. L. (1969). On how so little information controls so much behaviour. In *Towards a theoretical biology: 2. Sketches* (ed. C. H. Waddington) pp. 236–46. Biological Sciences and Edinburgh University Press.

Gregory, R. L. (1970). *The intelligent eye.* Weidenfeld and Nicolson, London. [Sets out the philosophy of this present book, including experiments added in this new edition.]

Gregory, R. L. (1974). Choosing a paradigm for perception. In *Handbook of perception.* Vol. 1, (ed. E. C. Cartarette and M. P. Freeman), Chapter 3. Academic Press, New York. [Challenges alternative paradigms with test questions.]

Gregory, R. L. (1980). Perceptions as hypotheses. *Philosophical Transactions of the Royal Society,* **B 290**, 181–97. [This is the central theme of the present book]

Hebb, D. O. (1949). *The organization of behaviour.* Chapman and Hall/Wiley, London. [Pioneering account of the brain as an inter-active net.]

Johnson-Laird, P. N. (1983). *Mental models.* Harvard University Press, Cambridge, Mass.

Kanizsa, G. (1979). *Essays on Gestalt perception.* Praeger, New York.

Koffka, K. (1935). *Principles of Gestalt psychology.* Harcourt Brace, New York.

Shannon, C. E. and Weaver, W. (1949). *The mathematical theory of communication.* University of Illinois Press, Urbana. [Essential account of information theory]

Watt, R. J. (1988). *Visual processessing: computational psychological, and cognitive research.* Lawrence Erlbaum, Hove. [Technical account in terms of physics and physiology.]

Winston, P. H. (1984). *Artificial intelligence.* Addison-Wesley, Reading, Mass.

2. Light

Bragg, W. (1962). *Universe of light.* Bell Clarke, London.

Feynman, R. (1985). *QED: The strange theory of light and matter.* Princeton University Press.

Hecht, S., Shlaer, S., and Pirenne, M. H. (1942). Energy, quanta, and vision. *Journal of General Physiology,* **25**, 819–40.

Jenkins, F. A. and White, H. E. (1957). *Fundamentals of optics* (3rd edn). McGraw Hill, New York.

Newton, I. (1730). *Opticks* (3rd edn). Dover Publications reprint 1952.

Ronchi, V. (1957). *Optics: the science of vision* (trans. E. Rosen). New York University Press. Dover Publications reprint 1991.

Ronchi, V. (1970). *The nature of light.* Heinemann. London. (Trans. of *Storia della Luce* 1939.)

3. Eye

Campbell, F. W. and Whiteside, T. C. D. (1950). Induced pupillary oscillations. *British Journal of Ophthalmology*, **34**, 180.

Dowling, J. E. (1987). *The retina—an approachable part of the brain*. Harvard University Press, Cambridge, Mass.

Gregory, R. L. (1986). See Naples and live: the scanning eye of Copilia. In *Odd perceptions*. Routledge, London.

Gregory, R. L. and Ramachandran, V. S. (1991). Perceptual filling in of artificially induced scotomas in human vision. *Nature*, **350**, 699–702. [Gives experimental evidence that filling-in is an active process—rather than the brain ignoring regions giving no information.]

Gregory, R. L. and Zangwill, O. L. (1963). The origin of the autokinetic effect. *Quarterly Journal of Experimental Psychology*, **15**, 4.

Grimes, J. (1996). Failure to detect changes in scenes across saccades. In *Perception* (ed. K. Akins) (Vancouver Studies in Cognitive Science, Vol. 5), pp. 89–110. Oxford University Press.

Hammond, J. H. (1981). *The camera obscura*. Adam Hilger, Bristol.

Harmon, L. D. (1974). The recognition of faces. In *Image, object and illusion*. Readings from Scientific American pp. 101–12. W. H. Freeman, San Francisco.

Howard, I. P. and Rogers, B. J. (1995). *Binocular vision and stereopsis*. Oxford University Press.

Julesz, B. (1971). *Foundations of cyclopean perception*. Chicago University Press.

Land, M. F. (1984). Molluscs. In *Photoreception and vision* (ed. M. A. Ali) pp. 699–725. Plenum, New York.

Ogle, K. N. (1950). *Researches in binocular vision*. Saunders, London.

Pirenne, M. H. (1948) *Vision and the eye*. Chapman and Hall, London.

Riggs, L. A., Ratliff, E., and Cornsweet, T. N. (1953). The disappearance of steadily fixated visual test objects. *Journal of the Optical Society of America*, **43**, 459.

Wade, N. J. (1983). *Brewster and Wheatstone on vision*. Academic Press, London.

Walls, G. L. (1963). *The vertebrate eye and its adaptive radiation*. Hafner, New York.

Wheatstone, C. (1838). Contributions to the physiology of vision, Part 1: On some remarkable and hitherto unobserved phenomena of binocular vision. *Philosophical Transactions of the Royal Society of London*, **128**, 371–94.

Yarbus, A. (1967). *Eye movements and vision*. Plenum, New York.

4. Brain

Anstis, S. M. and Gregory, R. L. (1964). The after-effect of seen motion: The role of retinal stimulation and eye movements. *Quarterly Journal of Experimental Psychology*, **17**, 173–4.

Biederman, I. (1987). Recognition-by-components: A theory of human image understanding. *Psychological Review*, **94**, 115–47. [This is somewhat different from David Marr's account, being based on human perception rather than AI]

Biederman, I. and Cooper, E. E. (1991). Priming contour-deleted images: evidence for intermediate representations in visual object recognition. *Cognitive Psychology*, **23**, 393–419.

Blakemore, C. (ed.) (1990). *Vision: coding and efficiency*. Cambridge University Press.

Ewert, P. H. (1930). A study of the effect of inverted retinal stimulation upon spatially coordinated behaviour. *Genetic Psychology Monograph*, **8**.

Gilbert, C. G. and Wiesel, T. (1992). Receptive field dynamics in adult primary visual cortex. *Nature*, **356**, 150–2. [The physiological basis of filling-in scotomas]

Gross, C. G., Rochar-Miranda, G. E., and Bender, D. B. (1972). Visual properties of neurons in the inferotemporal cortex of the macaque. *Journal of Neurophysiology*, **36**, 96–111.

Hubel, D. H. and Wiesel, T. (1959). Receptive fields of single neurons in the cat's striate cortex. *Journal of Physiology*, **148**, 574–91.

Hubel, D. H. and Wiesel, T. N. (1962). Receptive fields, binocular interaction and functional architecture of the cat's visual cortex. *Journal of Physiology*, **166**, 106–54.

Lettvin, J. Y., Maturana, H. R., McCulloch, W. S., and Pitts, W. H. (1959). What the frog's eye tells the frog's brain. *Proceedings of the Institute of Radio Engineers of New York*, **47**, 1940–51.

Livingstone, M. S. and Hubel, D. H. (1984). Anatomy and physiology of a colour system in the primate visual cortex. *Journal of Neuroscience*, **4**, 309–56.

Livingstone, M. S. and Hubel, D. H. (1987). Psychophysical evidence for separate channels for the perception of form, colour, movement and depth. *Journal of Neuroscience*, **7**, 3416–68.

Lu, C. and Fender, D. W. (1972). The interaction of colour and luminance in stereoscopic vision. *Investigative Ophthalmology*, **11**, 6, 482–9.

Marr, D. (1982). *Vision: A computational investigation into the human representation and processing of visual information*. W. H. Freeman, San Francisco.

Perrett, D. I., Rolls, E. T., and Caan, W. (1982). Visual neurons responsive to faces in the monkey temporal cortex. *Experimental Brain Research*, **47**, 329–42.

Perrett, D. I., Harries, A. J., Benson, P. J., Chitt, M., and Mistlin, A. J. (1990). Three stages in the classification of body movements by visual neurones. In *Images and understanding* (ed. H. B. Barlow, C. Blakemore, and M. Western-Smith), pp. 94–108. Cambridge University Press.

Perrett, D. I., Benson, P. J., Hietanen, J. K., Oram, M. W., and Dittrich, W. H. (1995). When is a face not a face? In *The artful eye* (ed. R. L. Gregory, J. Harris, P. Heard, and D. Rose), pp. 95–124. Oxford University Press.

Rose, S. (1973). *The conscious brain*. Weidenfeld and Nicolson, London.

Shepard, R. N. and Cooper, L. A. (1983). *Mental images and their transformations*. MIT Press, Cambridge, Mass.

Shepard, R. N. and Metzler, J. (1971). Mental rotation of three-dimensional objects. *Cognitive Psychology*, **3**, 701–3.

Stratton, G. M. (1897). Vision without inversion of the retinal image. *Psychological Review*, **4**, 341–60 and 463–81.

Stroop, J. R. (1935). Studies of interference in serial verbal reactions. *Journal of Experimental Psychology*. **18**, 643–62. [The Stroop effect is that colour names printed in non-appropriate colours are hard to read]

Thompson, P. (1980) Margaret Thatcher: A New Illusion. Perception, **9**, p. 483.

Treisman, A. M. and Schmidt, H. (1982). Illusory conjunction in the perception of objects. *Cognitive Psychology*, **14**, 107–141.

Wertheimer, M. (1938). Laws of organisation of perceptual forms. In *Source book of Gestalt psychology* (ed. W. H. Ellis), pp. 71–88. Routledge Kegan Paul, New York.

Winograd, Terry (1972). *Understanding natural language*. Edinburgh University Press, Edinburgh.

Young, J. Z. (1978). *Programs of the brain*. Oxford University Press.

Zeki, S. M. (1993). *A vision of the brain*. Blackwell, Oxford.

5.   Seeing brightness

Arden, G. B. and Wheale, R. A. (1954). Variations in the latent period of vision. *Proceedings of the Royal Society of London Series B*, **142**, 258–67.

Barlow, H. B. (1956). Retinal noise and absolute thresholds. *Journal of the Optical Society of America*, **46**, 634–9.

Hecht, S., Schlaer, S. and Pirenne, M. H. (1942). Energy quanta and vision. *Journal of General Physiology*, **25**, 819–40.

Rushton, W. A. H. and Campbell, F. W. (1954). Measurement of rhodopsin in the human eye. *Nature*, **174**, 1096–7.

6.   Seeing movement

Anstis, S. M. and Gregory, R. L. (1965). The after-effect of seen motion: the role of retinal stimulation and eye movements. *Quarterly Journal of Experimental Psychology*, **17**, 173–4.

Anstis, S. M. and Ramachandran, V. S. (1995). At the edge of movement. In *The artful eye* (eds. R. L. Gregory, J. Harris, P. F. Heard, and D. Rose), pp. 232–48. Oxford University Press.

Braddick, O. (1980). Low-level and high-level processes of apparent motion. In *The psychology of vision* (ed.) H. C. Longuet-Higgins and N. S. Sutherland), pp. 137–51. The Royal Society of London.

Braddick, O. J. (1995). The many faces of motion perception. In *The artful eye* (ed. R. L. Gregory, J. Harris, P. Heard, and D. Rose), p. 205–31. Oxford University Press.

Ceram (1965). *Archaeology of the cinema*. Thames & Hudson, London.

Duncker, K. (1938). Induced motion. In *Source book of Gestalt psychology* (ed. W. H. Ellis) 161–172. Routledge and Kegan Paul, London.

Gregory, R. L. and Zangwill, O. L. (1963). The origin of the auto-kinetic effect. *Quarterly Journal of Experimental Psychology*, **15**, 252–61.

Johansson, G. (1973). Visual perception of biological motion and a model of its analysis. *Perception and Psychophysics*, **14**, 201–11.

Johansson, G. (1975). Visual motion perception. *Scientific American*, **232**, 66–89.

Miles, T. R. and Miles, E. (1963). *Perception and causality*. Methuen, London. [On Michotte's experiments]

Ramachandran, V. S. and Anstis, S. M. (1986). The perception of apparent motion. *Scientific American*, **254**, 80–7.

Shopland, C. D. and Gregory, R. L. 1964). 'The effect of touch on a visually ambiguous three-dimensional figure'. *Quarterly Journal of Experimental Psychology*, **16**, 66–70.

Wohlemuth, A. (1911). On the after-effect of movement. *British Journal of Psychology Monograph*, **I**.

7. Seeing colours

Binney, R. C. (ed.) (1961). *Colour vision*. Van Nostrand, Netherlands. [This contains Thomas Young's classical paper 'On the theory of light and colours' as well as those of Helmholtz]

Gregory, R. L. (1977). Vision with isoluminant colour contrast, 1. A projection technique and observations. *Perception*, **6**, 113–9.

Land, E. H. (1977). The retinex theory of colour vision. *Scientific American*, **237**, 108–28.

Livingstone, M. S. and Hubel, D. H. (1988). Segregation of form, colour, movement and depth: Anatomy, physiology, and perception. *Science*, **240**, 740–9.

MacAdam, D. L. (1970). (ed.) *Sources of colour science*. MIT Press, Cambridge Mass.

Walsh, V. and Kulikowski, J. (1995). Seeing colour. In *The artful eye* (ed. R. L. Gregory, J. Harris, P. Heard, and D. Rose), pp. 268–78. Oxford University Press.

Zeki, S. M. (1990). Colour vision and functional specialization in the visual cortex. *Discussions in neuroscience*, Vol. 6. Elsevier, Amsterdam.

8. Learning to see

Adaptations

Atkinson, J. (1995). Through the eyes of an infant. In *The artful eye* (ed. R. L. Gregory, J. Harris, P. Heard, and D. Rose), pp. 141–56. Oxford University Press.

Blakemore, C. and Cooper, G. C. (1970). Development of the brain depends on the visual environment. *Nature,* **228,** 477–8.

Cartwright, B. A. and Collett, T. S. (1983). Landmark learning in bees: experiments and models. *Journal of Comparative Physiology,* **151,** 521–43.

Ewert, P. H. (1930). A study of the effects of inverted retinal stimulation upon spatially co-ordinated behaviour. *Genetics and Psychology Monographs,* **7,** 177–363.

Foley, J. P., Jr (1940). An experimental investigation of the effect of prolonged inversion of the visual field in the rhesus monkey. *Journal of Genetics and Psychology,* **56,** 21–55.

Gould, J. L. (1985). How bees remember flower shapes. *Science,* **227,** 192–4.

Held, R. and Hein, A. (1963). Movement-produced stimulation in the development of visually guided behaviour. *Journal of Comparative and Physiological Psychology,* **56,** 872–6.

Peterson, J. and Peterson, J. K. (1938). Does practice with inverting lenses make vision normal? *Psychological Monographs,* **50,** 12.

Rock, I. (1973). *Orientation and form.* Academic Press, New York.

Slater, A. M., Mattock, A., Brown, E., Burnham, D., and Young, A. W. (1991). Visual processing of stimulus compounds in new-born babies. *Perception,* **20,** 29–33.

Smith, K. U. and Smith, W. M. (1962). *Perception and motion: an analysis of space-structured behaviour.* Saunders, London.

Stratton, G. M. (1896). Some preliminary experiments on vision. *Psychological Reviews,* **3,** 611.

Stratton, G. M. (1897). Vision without inversion of the retinal image. *Psychological Reviews,* **4,** 341.

Blindness—and recovery

Gregory, R. L. and Wallace, J. G. (1963). *Recovery from early blindness: a case study.* (Heffers Cambridge). Reprinted in *Concepts and mechanisms of perception* (1974) (ed. R. L. Gregory), pp. 65–129. Duckworth, London.

Hull, J. M. (1991). *Touching the rock: an experience of blindness.* Pantheon, New York. [Personal account of going blind]

Riesen, A. H. (1947). The development of visual perception in man and chimpanzee. *Science,* **106,** 107–8.

Senden, von M. (1960). *Space and sight.* (trans. P. Heath). Methuen/Free Press, London. [Historical accounts of the early cases of removal of the lens for blindness due to cataracts]

Valvo, A. (1971). *Sight restoration and rehabilitation*. American Foundation for the Blind, New York.

Infants and young animals

Bower, T. G. R. (1972). Object perception in infants. *Perception*, **1**, 15–30.

Bremner, J. G. (1994). *Infancy*, (2nd edn). Blackwell, Oxford. [Sound general account]

Bruner, J. S. and Kowslowaski, B. (1972). Visually pre-adapted constituents of manipulatory action. *Perception*, **1**, 3–14.

Bryant, P. (1974). *Perception and understanding in young children*. Methuen, London.

Fantz, R. L. (1961). The origin of form perception. *Scientific American*, **204**, 66–72.

Gibson, E. J. and Walk, R. D. (1960). The visual cliff. *Scientific American*, **202**, 64–71.

Hess, E. H. (1956). Space perception in the chick. *Scientific American*, **195**, 71.

Piaget, J. (1956). *The construction of reality in the child*. Routledge and Kegan Paul, London.

Piaget, J. (1956). *The child's conception of space*. Routledge and Kegan Paul, London.

Piaget, J. (1967). *The child's conception of the world*. Littlefield Adams, Totowa, NJ.

Slater, M. and Bremner, G. (eds) (1989). *Infant development*. Lawrence Erlbaum, Hillsdale (USA).

Stern, N. D. (1985). *The interpersonal world of the infant*. Basic Books, New York.

Forgetting

Luria, A. R. (1970). *Traumatic aphasia*. Mouton, The Hague.

Sacks, O. (1985). *The man who mistook his wife for a hat*. Duckworth, London.

Shallice, T. (1991). *From neuropsychology to mental structure*. Cambridge, University Press.

Verny, T. (1981). *The secret life of the unborn child*. Summit Books, New York.

9. Realities of art

Seeing space

Ames, A., Jr (1951). Visual perception and the rotating trapezoid window. *Psychological Monographs* **7**, 1–32.

Cantril, H. (ed.), (1960). *The morning notes of Adelbert Ames, Jr*. Rutgers University Press, New Brunswick.

Ittelson, W. H. (1952). *The Ames demonstrations in perception*. Princeton University Press.

Representing space

Gombrich, E. H. (1959). *Art and illusion*. Phaidon, London.
Helmholtz, H., von. (1881). On the relation of optics to painting. In *Popular lectures on scientific subjects* (2nd series) (trans. E. Atkinson). Longmans Green, New York.
Kemp, M. (1990). *The science of art: optical themes in Western art from Brunelleschi to Seurat*. Yale University Press.
White, J. (1967). *The birth and rebirth of pictorial space*. Faber and Faber, London.

10.   Illusions

Physiology

Carpenter, R. H. S. and Blakemore, C. (1973). Interactions between orientations in human vision. *Experimental Brain Research*, **18**, 287–303.
Iversen, S. D. and Iversen, L. L. (1975). *Behavioural pharmacology*. Oxford University Press.
Oswald, I. (1962). *Sleeping and waking, physiology and psychology*. Elsevier, Amsterdam.
Penfield, W. and Roberts, L. (1959). *Speech and brain mechanisms*. Oxford University Press.

Cultural effects

Deregowski, J. B. (1973). Illusion and culture. In *Illusion in nature and art* (ed.) R. L. Gregory and E. Gombrich, Duckworth, London.
Segall, M. H., Campbell, T. D. and Herskovitz, M. J. (1966). *The influence of culture on visual perception*. Bobbs Merrill, New York.

Illusions

Bartlett, J. C. and Searcy, J. (1993). Inversion and configuration of faces *Cognitive Psychology*, **25**, 281–316.
Blakemore, C., Carpenter, R. H. S., and Georgeson, M. A. (1970). Lateral inhibition between orientation detectors in the human visual system. *Nature*, **228**, 37–9.
Ernst, B. (1986). *Adventures with impossible figures*. Tarquin. Diss (Norfolk, England).
Frisby, J. P. and Davies, I. R. L. (1970). Is the haptic Muller–Lyer a visual phenomenon? *Nature*, **231**, 5303.
Gregory, R. L. (1963). Distortion of space as inappropriate constancy scaling. *Nature*, **199**, 678–80. [The first statement of the 'inappropriate constancy' theory of cognitive distortion illusions]

Gregory, R. L. (1968). Perceptual illusions and brain models. *Proceedings of the Royal Society of London, Series* B, **171**, 179–296.

Gregory, R. L. (1972). Cognitive contours. *Nature*, **238**, 51–2. [Suggests that illusory surfaces are visual 'postulates' for explaining gaps]

Gregory, R. L. (1995). Brain-created visual motion: An illusion? *Proceedings of the Royal Society of London, Series* B, **260**, 167–8. [explains Op-Art 'jazzing' as due to fluctuations of accommodation]

Gregory, R. L. and Gombrich, E. H. (eds) (1973). *Illusion in nature and art.* Duckworth, London.

Gregory, R. L. and Harris, J. P. (1974). Illusory contours and stereo depth. *Perception & Psychophysics* vol. 15, 3, 411–16.

Gregory, R. L. and Harris, J. P. (1975). Illusion-destruction by appropriate scaling. *Perception*, **4**, 203–20.

Gregory, R. L. and Heard, P. (1972). Border locking and the café wall illusion. *Perception*, **8**, 365–80.

Hill, H. and Bruce, V. (1994). A comparison between the hollow-face and hollow-potato illusions. *Perception*, **22**, 887–97.

Kaufman, L. and Rock, I. (1962). The moon illusion. *Science*, **136**, 953–61. [Interesting measurements. Theory different from this book's, as accepts Ptolomy's account that apparent size is given simply by apparent distance]

Luckiesh, M. (1922). *Visual illusions.* (Dover Publications reprint, 1965).

Mackay, D. M. (1961). Interactive processes in visual perception. In *Sensory communication* (ed. W. A. Rosenblith). MIT Press and Wiley Cambridge, Mass. [Explains Op-Art 'jazzing' as information overload]

Nakayama, K., Shimojo, S., and Silverman, G. H. (1989). Stereoscopic depth: its relation to image segmentation, grouping and the recognition of occluded objects. *Perception*, **18**, 55–68.

Penrose, L. S. and Penrose, R. (1958). Impossible objects: a special type of illusion. *British Journal of Psychology*, **49**, 31. [First publication of impossible figures]

Petry, S. and Meyer, Gl. (1987). *The perception of illusory contours.* Springer-Verlag, New York. [Useful history and research papers]

Robinson, J. O. (1972). *The psychology of visual illusions.* Hutchinson, London. [Comprehensive history. Needs updating]

Tolansky, S. (1967). *Optical illusions.* Pergamon, New York.

Zeki, S. (1993). Going beyond the information given: The relation of illusory visual motion to brain activity. *Proceedings of the Royal Society of London, Series B*, **252**, 215–2. [Explains Op-Art 'jazzing' as brain-generated]

Constancy scaling

Holway, A. H. and Boring, E. G. (1941). Determinates of apparent visual size with distance varients. *American Journal of Psychology*, **54**, 21–37. [Authoritative study]

Thouless, R. H. (1931). Phenomenal regression to the real object. *British Journal of Psychology*, **21**, 339.

Thouless, R. H. (1932). Individual differences in phenomenal regression. *British Journal of Psychology* **22**, 216. [Classic experiments]

11.   Speculations

Crick, F. (1994). *The astonishing hypothesis: The scientific search for the soul.* Macmillan, London. [The joint discoverer of the structural significance of DNA for genetics, has spent over 10 years considering the brain. Here are his conclusions, written with characteristic verve.]

Darwin, C. (1872). *The expression of the emotions in man and animals.* John Murray, London. (University of Chicago Press reprints 1965). [This classic is a joy to read and its trenchant ideas have been confirmed many times over]

De la Mettrie, J. (1748). *L'homme machine* (trans. *Man a machine*, 1953). Open Court, La Salle, Illinois. [Classic statement that mind is brain-generated. The author lost his medical practice and had to leave Paris for advancing the 'astonishing hypothesis' which all brain scientists now believe.]

Dennett, D. (1991). *Consciousness explained.* Little, Brown, New York. [Insightful, fascinating. Attacks dualism and the 'Cartesian Theatre'. But much remains to be explained!]

Gardner, H. (1987). *The mind's new science: a history of the cognitive resolution.* Basic Books, New York.

Gregory, R. L. (1997). *Mirrors in mind.* Spektrum, Oxford; Freeman, San Francisco. [Account of perceptual puzzles, optics, history, and mythology of mirrors]

Jeannerod, M. (1997). *The cognitive neuroscience of action.* Blackwell, Oxford.

Milner, P. and Goodale, M. A. (1995). *The visual brain in action.* Oxford University Press.

Mishkin, M., Ungerleider, L. G., and Macko, K. O. (1983). Object vision and spatial vision: two cortical pathways. *Trends in Neuroscience,* **6**, 414–17.

Parker, S. T., Mitchell, R. W., and Boccia, M. L. (1994). *Self-awareness in animals and humans.* Cambridge University Press.

Penrose, R. (1989). *The emperor's new mind: concerning computers, minds, and the laws of physics.* Oxford University Press. [Outstanding physicist's view of properties of mind given by quantum principles]

Pinker, S. (ed.) (1984). *Visual cognition.* Bradford, Cambridge, Mass.

Polyshyn, Z. W. (1973). What the mind's eye tells the mind's brain. *Psychological Bulletin,* **80**, 1–24.

Weiskrantz, L. (1986). *Blindsight: A case study and its implications.* Oxford University Press. [New evidence for (limited) unconscious human vision and its implications for considering the normal role of consciousness]

Index

ablation 68
accommodation of lens 37–8
acuity 56
action potentials 68
active vision 9
activity in brain 81–3
adaptation
  to disturbed image 138–46
  to image 47
  perceptual 143
  proprioceptive 143
after-effects
  contingent 147, 149
  figural 144–5
after-images 57–8
  rotary 109
agnosia 168–9
algorithm, computer 80
Alhazen 34
ambiguities in vision 11, 53, 195, 205–8,
    227–31, 247–9
  perceptual 195
ambiguous objects 9
Ames
  Albert 185–6
  room 185–7
  window 158
Anableps 37–8
analogue computer 82–3
anatomy of eye 36–43
anomalascope 131–2
anomaly 130–2
apparent depth, measurement of 234
aqueous humour 25, 36
architecture 173–5, 177–9, 184–8, 192, 214
area striate 68–70, 75, 76
Aristotle 110
art
  cave art 170–1
  Chinese art 173–5
  Egyptian art 172–3
  Japanese art 173–5
  Op-Art 200–4
  oriental art 173–5
  perspectives in 171–84

  realities of 170–93
arthropod 25
artificial nets 82–3
astigmatism 145
Atkinson, Janette 164
autokinetic phenomenon 105–9

baby vision, see infant
Barbaro, Damielo 34
bee 134, 137
behaviour 8, 158, 160
  chicken 139
  monkey 139
  see also infant
behaviourism 3
Berkeley, Bishop George 107, 152, 153,
    161–2
Biederman, Irving 79–80
binocular cell 62
Blakeman, Colin 138, 217
bleaching of pigments 57, 86
blind spot 58–60
blindness 85
  colour 128, 130–3
  recovery from 151–8
blinking 44, 75
blue–green colour blindness 130
bottom-up signals 11, 206–8, 227–31, 249,
    251–2, 254
Bower, T. G. R. 163
Bradick, Oliver 164
brain activity 4, 67–83
  damage to 67
  imaging 81–3
  simulation 74–5
Brewster, David 190
brightness, seeing 84–97
Brown, G. C. 142
Brunelleschi, Filippo 175
Bruner, Jerome 163

café wall illusion 209–12, 216, 249
camera obscura 34–5, 176, 181
Canaletto, Antonio 177
cartoon 120

causality, illusion of 119–20
cave art 170–1
cell death in lenses 38
cephalopod 53
cerebral cortex 68, 75
Cézanne, Paul 79
Cheseldon, William 153
chicken behaviour 139
child development, *see* infant
Chinese art 173–5
cinema 116, 118–19
cochlea 37
cognitive psychology 4
colours 55, 71–2
    blindness 128, 130–3
    complementary 128
    constancy 133–4
    contrast of 87–92
    detection of 77–9
    red–green blindness 130
    receptors 122, 124
    seeing 121–35
    subtractive 128–9
    triangle 128–9
complementary colours 128
complexity of eyes 25
compound eye 25, 27–8, 32–3
computer algorithm 80
conceptual illusion 196–7
conditioned reflex 3
cones 54–5, 86–7
consciousness 252–5
constancy scaling 184–8, 222–37
contingent after-effects 147, 149
contrast of colour 87–92
copepod 29–30
Copilia 29–30, 32
cornea 36, 37, 44
corneal lens 27
corpus callosum 75
correcting vision 50–2
Craik, Kenneth 4
crystalline lens 37–9
cultural differences in vision 150–1
Curie, Pierre 212

Daphnia 29
dark–light adaptation 57, 85–7
Darwin, Charles 2, 24
decision-taking 5
deduction 246
depth vision 60–6, 177, 185, 192–3, 232–4
    cues 164–7, 217–27
    flipping 11, 64, 66, 207–8, 210, 231
    measuring 232, 234
    reversal 64–6, 190–1

Descartes, René 17–18, 37, 52, 151, 195–6, 222–4
deuteranopia 130–1
development of infant, *see* infant
digital computer 82–3
dim light 28, 57
disparity 61, 64
displaced images 138–43
displacement in time 145–6
distortion 144, 204, 208–12, 231, 247–9
    by absence of perspective 232
    empathy theories for 214
    eye movement theories for 214, 216
    feature detector theories for 216–17
    good-figure theory for 212, 214
    limited acuity theories for 216
    perspective 217–27
disturbed image, adaptation to 138–46
dreams 198–200
    hypnagogic 199
drugs 74, 199
Duncker, Carl 113
Dürer, Albrecht 177
dynamics 3–4

Egyptian art 172–3
Einstein, Albert 107
electromagnetic radiation 17–21
Emmert's law 225–6, 231
emotions 244–5
empathy theories for distortion 214
Erisman, T. 142
Euclid 14
Euclidian geometry 137
evolution 2
    of eye 24–33
Ewert, P. H. 142
Exner, Selig 29
eye 1–13, 21, 24–66
    anatomy 36–43
    complexity of 25
    compound 25, 27–8, 32–3
    evolution of 24–33
    movement 44–7, 164–5
        recording of 45, 101–2
        theories for distortion 214, 216
    muscles 44–5
    as optical instrument 34–5
    passive movement 102, 104–5
    pigments 124, 127–8
    primitive 26–7
    receptor 54–6
    scanning 29–30
    sensitivity 22, 92–6
    simple 25, 32–3
    voluntary movement 102, 104–5

eyeglasses 50–2
eye–head movement 100–4, 107, 109, 112

face recognition 74, 164, 166
Fantz, Robert 164–6
Faraday, Michael 253
feature detector 138
    theories for distortion 216–17
Fechner's paradox 88–9
fiction 195, 204, 239–43, 247–9
figural after-effects 144–5
figure–ground switching 9, 11, 249
fish vision 37–8
fixing retinal images 47–50
flicker 116, 118, 249
flipping 205–8, 228–30
    depth 11, 64, 66, 207–8, 210, 231
focusing of lens 37–8
    on retina 52–7
forgetting how to see 168–9
Foucault, Jean Bernard Léon 18
fovea 44, 55, 98
Freud, Sigmund 5, 199
Frisby, John 235

Gauss, Karl 137
geometric perspective 175–7
geon 80
Gestalt therapy 3–4, 5, 6, 106, 113, 118, 142,
    208, 212
Gibson, Eleanor 164–6
Gibson, James J. 9
Giotto 184
Goodale, Mervyn 160
good-figure theory for distortion 212–14
Greek philosophers 1, 14
grey vision 55
Grimes, James 47–9
Gross, Charles 73
grouping of patterns 6

hallucinations 198–200
Harris, John 233
Hebb, Donald 82
Hecht, S. 23
Hein, Alan 143–4
Held, Richard 143–4
Helmholtz, Hermann von 2, 4–5, 42, 69,
    102–3, 122, 126, 216
holism 3–4
hollow mask 64, 66, 207–8, 210
homunculus 53, 68, 70
Hooke, Robert 121
horseshoe crab, *see Limulus*
Hubel, David 75–7, 78
hue discrimination 126–7

Huxley, Aldous 199
Huygens, Christian 14, 16, 18
hypnagogic dreaming 199
hypothesis of perception 9–13

iconoscope 65
illusion 112, 194–243, 246
    causality of 119–20
    cognitive 197
    conceptual 196–7
    definition 194–5
    destruction of 233–5
    of moon 230–1
    Muller–Lyer arrows 151, 214, 216,
        219–21, 226, 233, 234, 237, 248, 249
    perceptual 196–7
    perspective 212–27
    physical 247–8
    Poggendorff 235
    Ponzo 218–19, 226, 231, 249
    size–weight 197–8, 249
    Titchner 158, 160
    waterfall effect 109–12
illusion-destruction 233–5
illusory triangle 239–40
image
    adaptation to 47, 138–46
    displaced 138–43
    distortions 144, 204, 208–12, 231, 247–9
    disturbed 138–46
    stabilizing 47–50
image–retina movement 99–104, 109, 112,
    118
imaging the brain 81–3
impossible triangle 235, 238–9, 240, 248,
    249
inappropriate constancy-scaling theory
    225–7
indirect vision 9
induced movement 113–14
induction 246
infant
    boredom 163
    development of sight 136–7, 156, 158,
        250
    permanence of objects 161–2
    prediction 162
    preferences 163
    surprise 162–3
    vision 6, 161–9
    visual cliff 164–7
inflow theory 102, 104
innate knowledge 136–8
insect 25, 27–9
instinct 136–7
intelligence, machine 75

inverted vision 6, 52–3, 139–43, 145
inverting goggles 139–43
iodopsin 57
iris 37, 39–41
isomorphic brain traces 5
Ittleson, W. H. 226

James–Lange theory 245, 255
Japanese art 173–5
Julesz
   bela 62–4
   technique 217, 62–4

Kaffka, Kurt 106
Kanizsa
   Gaetano 239–42
   triangle 239–42, 248, 249
Kepler, Johannes 35, 52, 53
Kohler, Ivo 142, 146–7
Kohler, Wolfgang 4

Land, Edwin 133–5
Land, Michael 29–30
lateral geniculate nucleus 78
law of
   qualities 97
   refraction 17–18
learned knowledge 136–8
learning how to see 136–9, 153–8
left–right reversal 5, 139, 142–3, 145
lens 14, 24–30, 36, 58
   accommodation of 37–8
   cell death in 38
   change in shape 38–9
   corneal 27
   crystalline 37–9
   cylinder 27, 29
   focusing of 37–8
Lettvin, J. Y. 99
light 14–23
   dim 28, 57
   nature of 18
   response to 25
   sensitivity 22, 92–6
   speed of 17, 20
   wavelength of 20–1, 84, 197
light–dark adaptation 57, 85–7
limited acuity theories for distortion 216
limpet 25
Limulus 92–3
Lipps, Theodore 214
Livingstone, Margaret 78
Locke, John 152, 153
luminosity curve 89

Mach, Ernst 105, 206, 209

Mach's bands 55
machine
   intelligence 75
   vision 79–80, 83
McCollough, Celeste 147, 149
MacKay, Donald 200–2
magno cell 78
mammal 53
Marr, David 79, 80
Maxwell, James Clark 122
measurement of visual depth 232, 234
mesopic vision 55
Michotte, Albert 119
Mill, John Stuart 161
Milner, David 160
Minsky, Marvin 82
mirror reflection 195, 249
moiré pattern 200, 202, 249
Molyneux, William 152, 153
monkey behaviour 139
moon illusion 230–1
motion parallax 114–15, 227
movement 100–4, 107, 109, 112
   detection 98–120
   of eye 44–7, 164–5
     recording of 45, 101–2
   eye–head 100–1
   image retina 99–101
   induced 113–14
   passive eye 102, 104–5
   perception of 112–14
   phi 118–19
   receptors 99–126, 200–4
   recording of 45
   relativity of 112–14
   seeing of 98–210
   signals of 116
   velocity 99–100
   of voluntary eye 102, 104–5
Muller, Johannes 70, 97
Muller–Lyer arrows illusion 151, 214,
   216, 219–21, 226, 233, 234, 237,
   248, 249
myelin sheath 69

Nautilus 25
Necker cube 188, 205, 227, 249
negative perspective 173–4, 185–6
nerve
   cell 68–9
   electrical
     activity 92–3
     conduction 73
   impulse speed 70–1
   optic fibre 75, 93
   sensory 97

nervous system 122
neural net 82–3
Newton, Sir Isaac 14, 18–19, 121, 195–7
NMR 68, 81
noise
  activity 92–5
  neural 94–5
nuclear magnetic resonance, *see* NMR

object
  changing 11
  constancy 162
  recognition 161
occipital region 69
occlusion 193
Offray de la Metrie, Julien 253
ommatidium 27–8, 29
Op-Art 200–4
ophthalmoscope 42–3
optic
  chiasma 75–6
  nerve fibre 75, 93
optical instrument (eye) 34–5
organ of Corti 37
organizing patterns 6
oriental art 173–5
outflow theory 102–4

painting 170–1, 249
Panum, P. L. 62
Papert, Seymour 82
paradox 112, 114–15, 204, 228–30, 235–9,
    247–9
  perceptual 195
parvo cells 78
passive
  eye movement 102, 104–5
  vision 2
Pasterson, J. K. 142
pattern
  grouping 6
  moiré 200, 202, 249
  organizing 6
  recognition 4, 6, 161
Pavlov, Ivan 3
Penfield, Wilder 198
Penrose, Lionel 235
Penrose, Roger 235, 253–4
perception 55, 158–60, 177–84, 228–30,
    245–6
  adaptation to 143
  consistency of 149
  hypothesis 9–13
  of movement 112–14
  paradigm of 3–13
  velocity 112

perceptual
  ambiguities 105
  illusion 196–7
  paradox 195
  scaling 172
permanence of objects 161–2
Perrett, David 73
perspective
  absence of 232
  ambiguities of 171–84
  in art 171–84
  distortion 217–27
  geometric 175–7
  illusion 212–27
  negative 173–4, 185–6
PET 68, 81, 252
Pfister, M. H. 139
phenomenal phenomena 248–50
  classification of 204–12
phi movement 118–19
photographs 184–5, 134
photon 22
photopic vision 55
physics of seeing 23
Piaget, Jean 160, 161–2
pictures 170–1, 249
pigments, eye 124, 127–8
pinhole
  camera 34
  image 14–15
Pirenne, M. H. 23
Plato 14
Poggendorff illusion 235
pointillism 128
Poisson distribution 23
Ponzo illusion 218–19, 226, 231,
    249
Porta, Giovanni Batista della 34
positron emission tomography,
    see PET
pregnance 3–4, 212–14
primitive eye 26–7
principal colours 123–7
printing 127–8
prism 18, 20
problem solving 3–5
protanopia 130–1
pseudoscope 64–6
Pulfrich pendulum effect 90–2
pupil 42–4, 89
Purkinje
  Johannes Evangelista 89
  shift 90

qualia 251, 253–5
quantum 22–3

range-finder convergence 60–1, 64
Rayleigh, Lord 131
realities of art 170–93
receptor
  colour 122, 124
  in eye 54–6
  for movement 99–126, 200–4
recognition of
  face 74, 164, 166
  object 161
  pattern 4, 6, 161
recording eye movements 45
red-eye, cause of 42
red–green colour blindness 130
reflex 8
  of blinking 44
  conditioned 3
  innate 3
refraction 17
refractive index 25, 37
Reisen, A. H. 151
relativity of movement 112–14
restoration of sight 151–8
retina
  fixing of images 47–50
  focusing 52–7
  seeing your own 58
retinal rivalry 61–2, 249
Reutersvärd, Oscar 235
reversal of vision 64–6, 190–1
rhodopsin 57, 86
right–left reversal 5, 139, 142–3, 145
Riley, Bridget 200–1, 203
robotics 79
rods 54–5, 86
Roemer, Olaus 17
Rogers, Brian 114
Rosenblatt, Frank 82
Rushton, W. A. H. 86

saccades 44, 47–9
Sacks, Oliver 157, 168–9
scanning eye 29–30
Scheiner, C. 52
Schlaer, S. 23
Schumann, F. 239, 241
scotoma 58–60, 249
scotopic vision 55
seeing
  brightness 84–97
  cause 119–20
  colour 121–35
  movement 98–210
  primitive 98
Sejnowski, Terence 83
semantics 5, 120, 251

sensitivity
  of eye 22, 92–6
  to light 22, 92–6
sensory nerves 97
servo-control system 40–1
shading 189–92
shadow 189–92
shape constancy 225
Sherrington, Sir Charles 102
side-ways rules 227, 247, 251
sight, restoration of 151–8
signals of movement 116
simple eye 25, 32–3
sine law 17–18
size constancy scaling 184–8, 222–7
size–weight illusion 197–8, 249
Skinner, B. F. 3
Smith, K. V. 145–6
Snell, Willebrod von Roijen 17
spectacles 50–2
spectrum 18, 20, 89, 126–7
speed of light 17, 20
Sperry, R. W. 139
stabilizing image 47–50
stereo projection 115
stereoscope 61, 66, 242
  Wheatstone 61
stereoscopic vision 60–6, 138, 163, 217, 232
Stratton, G. M. 139–42
stress 44
striate cortex 68–9, 70
superior colliculus 75
surface cortex 68
symbols 244
syntax 5, 120, 251

telestereoscope 65
television 116, 118–19, 145
texture 191–2
theory
  inflow 102, 104
  James–Lange 245, 255
Thompson effect 74
Thoules, Robert 223, 225
three-dimensional vision 60–6, 235–7
threshold intensity 95
time displacement 145–6, 148
Titchner illusion 158, 160
top-down knowledge 11, 66, 72, 206,
  208, 227–32, 240, 247, 249, 251–2,
  254
touch 6, 30, 53, 143, 151, 235
trapezoid window 188
Treisman, Anne 71–2
trilobite 25, 27
tritanopia 130–1

tropic responses 7

upside down vision 5–6, 52–3, 139–42, 145

Valvo, Alberto 157
velocity
  measurement 99–100
  perception of 112
Vermear, Jan 177
Vinci, Leonardo da 176–7
visible light 21
vision 1–13
  active 9
  agnosia 168–9
  cliff experiment 164–7
  correcting 50–2
  cultural differences in 150–1
  depth 60–6, 177, 185, 192–3, 232–4
  fish 37–8
  grey 55
  indirect 9
  infant 6, 161–9
  inverted 6, 52–3, 139–43, 145
  machine 79–80, 83
  mesopic 55
  passive 2
  photopic 55
  postulates 242

psychology of 9
  reversal 64–6, 190–1
  scotopic 55
  upside down 5–6, 52–3, 139–42, 145
vitreous humour 52
voluntary eye movements 102, 104–5

Wallace, Jean 153
wandering light 105–9
waterfall effect 109–12
Watson, John Broadus 3
wavelengths of light 20–1, 84, 197
Weber's law 94–5
Western Renaissance 175–84
Wheatstone
  Sir Charles 61
  stereoscope 61
Wiesel, Torstin 75–7
Winograd, Terry 79
Wright, W. D. 126

Yarbus, Alfred 44, 46
yellow, perceptin of 124, 131
Young, Thomas 122–6, 134

Zeki, Semir 134
zonula 38

Acknowledgements

My interest in perception started with the teaching of Professor Sir Frederic Bartlett, FRS, and was encouraged by Professor O. L. Zangwill, at Cambridge University.

I would like to thank particularly Dr Stuart Anstis, and my other colleagues and students who have discussed the problems of this book with me, pointed out errors, and helped with experiments. I have benefited by the personal generosity of many people in the United States, especially: the late H.-L. Teuber, the late Warren McCulloch, and F. Nowell Jones. The book was largely written during a visit to Professor Jones's department at UCLA.

I would like to thank Mrs Audrey Besterman and Miss Mary Waldron for drawing the majority of the diagrams and the Illustrations Research Service, London, for collecting the colour plates used.

Acknowledgement is due to the following for illustrations: **1.3** C. E. Osgood and Oxford University Press; **2.3** The British Museum; **2.5, 7.1** The Royal Society; **2.6** The Bodleian Library; **3.1** G. L. Walls and *Cranbrook Institute of Science Bulletin*; **3.2** M. Rudwick; **3.3** V. B. Wigglesworth, Methuen & Co., Ltd, and John Wiley and Sons, Inc.; **3.4, 3.5** R. L. Gregory, H. E. Ross, N. Moray and *Nature*; **3.9** redrawn from J. G. Sivak (1976) *Vision Research*, Vol. 16, pp. 531–4, copyright 1976 Elsevier Science Ltd., The Boulevard, Langford Lane, Kidlington OX5 1GB, UK; **3.10** T. C. Ruch, J. F. Fulton, and W. B. Saunders Co.; **3.14** R. M. Pritchard and *Quarterly Journal of Experimental Psychology*; **4.2** W. Penfield, T. Rasmussen and The Macmillan Co., New York; **4.4** The British Broadcasting Corporation; **4.7, 4.8** D. H. Hubel, T. H. Wiesel and the *Journal of Physiology*; **5.6** H. K. Hartline and Academic Press, Inc.; **6.2, 10.19** The Mansell Collection; **6.4** R. L. Gregory, O. L. Zangwill, and *Quarterly Journal of Experimental Psychology*; **6.9** A. Michotte, Methuen & Co. Ltd., and Basic Books, Inc.; **7.3** W. D. Wright and Henry Kimpton; **7.4** S. Hecht, C. Murchisson, and Clark University Press; **8.4, 8.5** K. U. and W. M. Smith and W. B. Smith and W. B. Saunders Co.; **8.6** I. Kohler and Scientific American; **8.7** J. Allen Cash; **8.8, 8.9** R. L. Gregory, J. G. Wallace, and *Experimental Psychology*

Society; **8.11** R. L. Franz and *Scientific American*, photo by David Linton; **8.13** William Vandivert and *Scientific American*; **9.3** John Freeman; **9.5** Bibliothèque de l'Institut de France; **9.6** by courtesy of the Trustees of the National Gallery, London; **9.7** Drawings Collection, Royal Institute of British Architects; **9.10**, **9.11**, **9.12** *Punch*; **9.13** Eastern Daily Press, Norwich; **9.16**, **10.15** Derrick Witty; **9.18** J. J. Gibson, Allen & Unwin Ltd., and Houghton Mifflin Company; **10.14** Rania Massourides.

R.L.G.

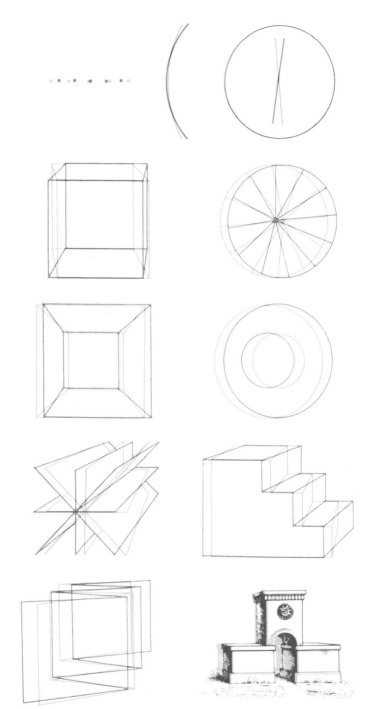

Plate 1 Wheatstone's drawings reproduced as red–green anaglyphs. (See pages 62–3.)

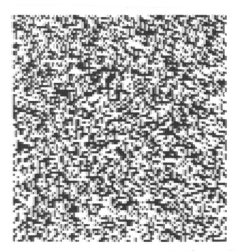

Plate 2 Julesz figure. Dot patterns, computer-generated with a small sideways shift of each dot in a central region of one picture, which in a stereoscope appears above the rest of the random pattern. This is a remarkably useful technique. For example, distortion illusion figures can be presented as dots shared between the eyes, so that neither eye alone can see the figure, though it is seen by the brain combining (fusing) the two sets of dots. If the distortion is still there, its origin must be 'central' in the brain, rather than in the retinas. (See page 64.) (From Julesz, *Foundations of cyclopean perception* (1971).)

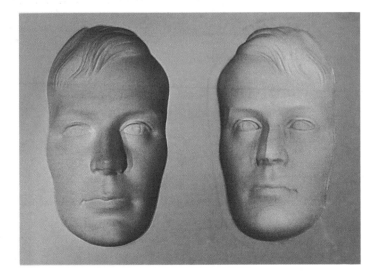

Plate 3 The hollow mask in stereo. The face on the left is normal—sticking out. The face on the right is actually hollow (cf. Figure 10.11). Does stereo information from the two eyes make the face on the right appear truly hollow? Here bottom-up stereo is pitted against top-down knowledge of faces. Which wins? The actually hollow face on the right appears convex like a normal face—in spite of the stereo information from your eyes. (It is better to investigate this with an actual hollow mask, viewed from different distances with both eyes.) (See pages 207–8, 210.)

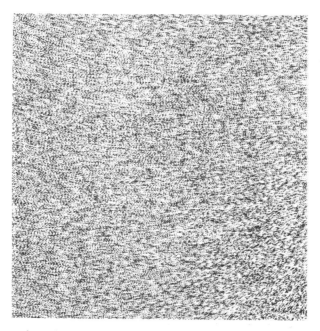

Plate 4 For each eye this is a random dot pattern. When the patterns are combined by the brain, a figure appears—the Muller–Lyer 'arrows', or 'corners' are seen (see Figure 10.16). This appears distorted, as usual for the Muller–Lyer illusion. So the origin of the distortion must be in the brain (following fusion of the eyes' images), not in the retinas. This makes a cognitive explanation probable. (See page 219.) (From Julesz, *Foundations of cyclopean perception* (1971).)

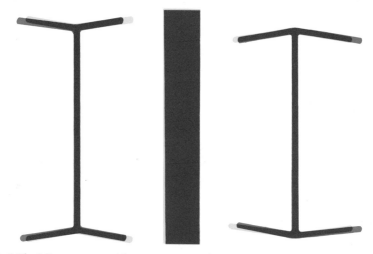

Plate 5 The 3-D appearance of the wire corners of the Muller–Lyer illusion. At the appropriate distance the distortion entirely disappears—when they are seen as corners rather than flat. This confirms that the usual distortion is due to inappropriate size-scaling, set by depth features on a flat surface. (See page 234.)

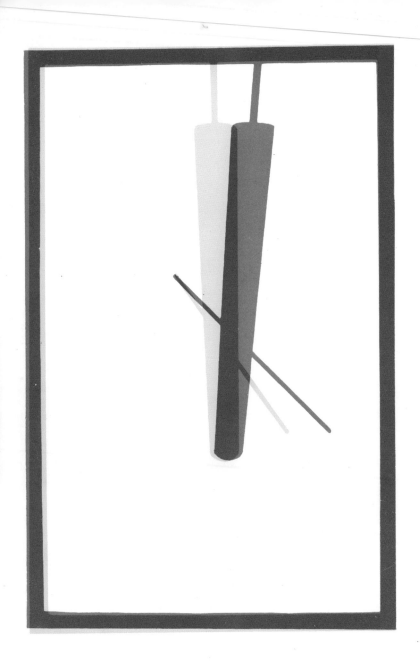

Plate 6 Poggendorff illusion in 3-D. When the oblique line is seen tilted in depth, the illusion disappears. It is now seen—as it is—as a straight line with no displacement. (See Fig. 10.13, page 213.)